Mechanical Engineering Series

The Mechanical Engineering Series presents advanced level treatment of topics on the cutting edge of mechanical engineering. Designed for use by students, researchers and practicing engineers, the series presents modern developments in mechanical engineering and its innovative applications in applied mechanics, bioengineering, dynamic systems and control, energy, energy conversion and energy systems, fluid mechanics and fluid machinery, heat and mass transfer, manufacturing science and technology, mechanical design, mechanics of materials, micro- and nano-science technology, thermal physics, tribology, and vibration and acoustics. The series features graduate-level texts, professional books, and research monographs in key engineering science concentrations.

More information about this series at https://link.springer.com/bookseries/1161

Chander Prakash • Sunpreet Singh
Seeram Ramakrishna
Editors

Additive, Subtractive, and Hybrid Technologies

Recent Innovations in Manufacturing

Springer

Editors
Chander Prakash
School of Mechanical Engineering
Lovely Professional University
Punjab, India

Sunpreet Singh
National University of Singapore
Singapore, Singapore

Seeram Ramakrishna
National University of Singapore
Singapore, Singapore

ISSN 0941-5122 ISSN 2192-063X (electronic)
Mechanical Engineering Series
ISBN 978-3-030-99568-3 ISBN 978-3-030-99569-0 (eBook)
https://doi.org/10.1007/978-3-030-99569-0

This Springer imprint is published by the registered company Springer Nature Switzerland AG
The registered company address is: Gewerbestrasse 11, 6330 Cham, Switzerland

Preface

The proposed book *Additive, Subtractive, and Hybrid Technologies* offers the reader with comprehensive insights of the recent research breakthroughs in the additive, subtractive, and hybrid technologies to provide incomparable design and manufacturing independences, as well as to upgrade the product performance characteristics through various possible feasible collaborations between the additive and subtractive technologies. Indeed, the intrinsic benefits and limitations of both additive and subtractive manufacturing technologies could be overcome through their much-appreciated hybridizations. This edited book will cover the recent innovations in the additive and subtractive manufacturing technologies. However, a special emphasis has been made on the use of both additive and manufacturing technologies in a one-after-another or simultaneous approach. The editorial team members of the proposed book are highly motivated that the interested readerships of the book will receive profound information on additive, subtractive, and hybrid manufacturing technologies.

Punjab, India — Chander Prakash
Singapore, Singapore — Sunpreet Singh
Singapore, Singapore — Seeram Ramakrishna

Contents

Contributors

Wasim Ashraf Department of Applied Mechanics, Motilal Nehru National Institute of Technology Allahabad, Allahabad, Uttar Pradesh, India

Vishvesh J. Badheka School of Technology, Mechanical Engineering Department, Pandit Deendayal Petroleum University, Gandhinagar, Gujarat, India

Ajay Batish Department of mechanical Engineering, Thapar Institute of Engineering and Technology, Patiala, Punjab, India

K. S. Boparai Department of Mechanical Engineering, MRS Punjab Technical University Bathinda, Bathinda, Punjab, India

G. S. Brar Department of Mechanical Engineering, National Institute of Technology, Srinagar, Uttarakhand, India

Suman Chatterjee Department of Mechanical Engineering, National Institute of Technology Rourkela, Rourkela, Odisha, India

K. Chawla School of Mechanical Engineering, Lovely Professional University, Phagwara, Punjab, India

Vijay Kumar Dwivedi Department of Mechanical Engineering, IET, GLA University, Mathura, Uttar Pradesh, India

N. Giri Department of Mechanical Engineering, National Institute of Technology, Srinagar, Uttarakhand, India

Anas Islam Department of Mechanical Engineering, IET, GLA University, Mathura, Uttar Pradesh, India

Mohd Yunus Khan Department of Mechanical Engineering, National Institute of Technical Teachers Training and Research, Chandigarh, India

Pawan Kumar Department of Physics, University Institute of Sciences, Chandigarh University, Gharuan, Punjab, India

Rakesh Kumar Department of Mechanical Engineering, Manipal University Jaipur, Jaipur, Rajasthan, India

Raman Kumar Department of Mechanical Engineering, Guru Nanak Dev Engineering College, Ludhiana, Punjab, India

Ranvijay Kumar Department of Mechanical Engineering, University Centre for Research and Development, Chandigarh University, Gharuan, Punjab, India

Sudhir Kumar Department of Mechanical Engineering, Thapar Institute of Engineering and Technology, Patiala, Punjab, India

Siba Sankar Mahapatra Department of Mechanical Engineering, National Institute of Technology Rourkela, Rourkela, Odisha, India

Ashutosh Mishra Department of Applied Mechanics, Motilal Nehru National Institute of Technology Allahabad, Allahabad, Uttar Pradesh, India

Arpan Mondal Department of Mechanical Engineering, National Institute of Technical Teachers Training and Research, Kolkata, West Bengal, India

B. S. Pabla Department of Mechanical Engineering, National Institute of Technical Teachers Training and Research, Chandigarh, India

Pulak M. Pandey Department of Mechanical Engineering, Indian Institute of Technology Delhi, New Delhi, India

Swastik Pradhan Department of Industrial Engineering, Lovely Professional University, Phagwara, Punjab, India

Chander Prakash Department of Industrial Engineering, Lovely Professional University, Phagwara, Punjab, India

P. Sudhakar Rao Department of Mechanical Engineering, National Institute of Technical Teachers Training and Research, Chandigarh, India

B. N. Sahoo Department of Mechanical Engineering, Sardar Vallabhbhai National Institute of Technology, Surat, Gujarat, India

Anshuman Kumar Sahu Department of Mechanical Engineering, National Institute of Technology Rourkela, Rourkela, Odisha, India

Supriya Sahu Department of Mechanical Engineering, IGIT, Sarang, Odisha, India

A. S. Shahi Department of Mechanical Engineering, Sant Longowal Institute of Engineering and Technology, Sangrur, Punjab, India

Arun Dutt Sharma Department of Mechanical Engineering, National Institute of Technical Teachers Training and Research, Chandigarh, India

Pawan Sharma Department of Mechanical Engineering, Sardar Vallabhbhai National Institute of Technology, Surat, Gujarat, India

Abhijeet Singh Department of Mechanical Engineering, Guru Nanak Dev Engineering College, Ludhiana, Punjab, India

Gurminder Singh SIMAP Lab, Université Grenoble Alpes, Gières, France

J. Singh School of Mechanical Engineering, Lovely Professional University, Phagwara, India

R. Singh Department of Mechanical Engineering, National Institute of Technical Teachers Training and Research, Chandigarh, India

Rupinder Singh Department of Mechanical Engineering, National Institute of Technical Teachers Training and Research, Chandigarh, India

Sukhinderpal Singh Department of Mechanical Engineering, Guru Nanak Dev Engineering College, Ludhiana, Punjab, India

T. P. Singh Department of mechanical Engineering, Thapar Institute of Engineering and Technology, Patiala, Punjab, India

Jasin Kumar Tarai Department of Mechanical Engineering, IGIT, Sarang, Odisha, India

Abhishek Kumar Tiwari Department of Applied Mechanics, Motilal Nehru National Institute of Technology Allahabad, Allahabad, Uttar Pradesh, India

Girish C. Verma Department of Mechanical Engineering, Indian Institute of Technology Indore, Indore, Madhya Pradesh, India

Experimental Study on Machinability of AISI 4340 Steel During Hard Turning by CBN Tool

Jasin Kumar Tarai, Supriya Sahu, Swastik Pradhan, Chander Prakash, Anshuman Kumar Sahu, and Siba Sankar Mahapatra

1 Introduction

In industries, the main target is to manufacture with low cost and high accuracy with less machining time. Turning is one of the commonly used machining processes for the cutting operation and specially for the finishing of components. In this process, the essential task is the selection of cutting parameters to accomplish high cutting performance. The selection of cutting parameters reflects the surface quality and dimensional accurateness of the machined components, which evaluate the quality of the product. Surface quality of finished product in turning is one of the major quality characteristics. Surface qualities of product quantify the technical quality of product as well as influence the cost of the product. The selection of the process parameters also affect the cutting force applied on the material surface during machining. The application cutting force on the work piece material while machining also influences the surface quality of the product [1, 2]. Recently, hard turning process is gaining popularity in manufacturing industries over the slow and cost-intensive grinding process due to less consumption of time, cost, and energy. Hard turning is extensively used in finishing operation and machining of work piece material having hardness greater than 45 HRC is called as hard turning. After the commercialization of cubic boron nitride (CBN) tools, hard turning extended more

J. K. Tarai · S. Sahu
Department of Mechanical Engineering, IGIT, Sarang, Odisha, India

S. Pradhan · C. Prakash
Department of Industrial Engineering, Lovely Professional University, Phagwara, Punjab, India

A. K. Sahu (✉) · S. S. Mahapatra
Department of Mechanical Engineering, National Institute of Technology Rourkela, Rourkela, Odisha, India

C. Prakash et al. (eds.), *Additive, Subtractive, and Hybrid Technologies*, Mechanical Engineering Series, https://doi.org/10.1007/978-3-030-99569-0_1

attention for the machining hard materials in comparison with traditional grinding [3]. During finish hard turning operation, complex and mutual interactions are produced between tool and work piece at the contact surfaces. Subsequently, significant cutting forces and extreme tribological conditions developed a dry severe cutting edge, which causes plastic deformation of the cutting edge. Another important characteristic of the machined surface is their roughness and its indexes. The roughness has more significance during the utilization of final product. The researchers have emphasized for a long time to predict these surface roughness indexes for a given process under the specific cutting environments. The use of cutting fluid during machining process creates more occupational risks. So, it is suggested to study machinability in dry conditions. It is suggested to machine hard materials in dry environment to perform at greater cutting speed which effects higher tool tip temperature and softens the work piece material. The advantages of hard machining are decease in machining cost, time saving, improve surface quality, and elimination of deformities due to temperature. The application of hard machining can be improved by the utilization of optimization techniques that helps the manufactures for taking decision in multi-objective quality characteristics for better machining [4, 5].

Many literatures in machining operation have suggested the effect of different tool materials and cutting parameters on tool life and machined surface characteristics during machining by taking different work piece material. Stakhniv and Devin have studied the influence wear of CBN-based composite tool insert during finish turning of hardened steel on tool vibration. Here, they have used CBN/Si3N4 insert during turning process and studied the effect of tool vibration on surface roughness. They found that the rise of tool wear up to its critical value has no substantial effect on the tool vibration, while above critical value of tool wear, there occurs a significant increment in tool vibration with increase in machined surface roughness [6]. Gangopadhyay et al. have analyzed the effect of cutting speed cutting tool surface properties on the formation of build-up-edge (BUE) during dry turning of aluminum alloy by using different cutting tools like uncoated K10, PVD TiN-coated K10, CVD diamond-coated K10, and polycrystalline diamond (PCD). Uncoated K10 and PVD TiN-coated tools are unsuccessful in preventing of BUE. While, CVD diamond-coated and PCD tools have brought down the BUE significantly at machining with lower value of cutting speed [7]. Bhusan et al. have performed the machining of 7075 aluminum alloy-SiC metal matrix composite by taking tungsten carbide and polycrystalline diamond (PCD) tool inserts. The surface roughness of the machined surface by carbide tool is found to be lower as compared to PCD at same parametric conditions. But, flank wear of the tungsten carbide tool is found to be more as compared to PCD at same parametric conditions [8]. Che-Haron has performed the dry machining of Ti-6Al-2Sn-4Zr-6Mo alloy by using uncoated cemented carbide tools. The tool inserts having finer grain size have shown longer tool life. The tool wear occurred because of flank face wear and extreme chipping action on the flank edge. With increase in cutting time, average flank wear increases. And by increase in feed rate and cutting speed, tool life decreases. The microstructure of the machined surface changes by increase of micro-hardness of the top white layer [9].

Nordgren et al. have analyzed the plastic deformation of cemented carbide tools by finite element model (FEM) during turning of quenched and tempered AISI 4340 steel and compared it with the actual experimental result of plastic deformation [10]. Ramesh et al. have studied the influence of process parameters on the surface roughness of machined part during turning of titanium alloy by using CVD-TiN-TiCN-Al_2O_3-TiN-coated carbide. They have found that feed has highest effect towards the surface roughness [11]. Jagadish and Samual have investigated the micro turning process of titanium alloy by using TiN/AlTiN-coated carbide tool. They have studied the effect of feed, speed, and depth of cut on cutting force and surface roughness [12]. Upadhyay et al. have analyzed the influence cutting parameters and amplitude of vibration on surface roughness during turning of Ti-6Al-4V alloy [13]. Similarly, Khan and Maity have performed the turning of titanium (grade 2) by taking cryogenically treated inserts and study the effect speed, feed rate, and depth of cut on cutting force, surface roughness, machining temperature, and material removal rate. Cutting force, machining temperature, and surface roughness are found to be less during machining by cryogenically treated tool [14]. Again, Khan and Maity have studied the effect of cutting speed and cooling method like dry cutting, flood cooling and minimum quantity lubrication using carbide inserts during turning of titanium (grade 2). They have found that the use of minimum quantity lubrication has given superior machining performances [15]. Parida and Maity have compared the machinability of Inconel 718, Inconel 625 and Monel 400 at room and hot temperature of 300 °C and 600 °C during turning operation. They have found that there is substantial decrease of cutting force, tool wear, chatter formation, and surface roughness with rise in tool life and chip tool contact length during hot machining for all the three work pieces [16]. Pattnaik et al. have studied the performance of different tool inserts like WC, WC + TiN, WC + Ti (C, N) + Al_2O_3, polycrystalline diamond (PCD), and cemented carbide (K-10) insert during dry machining of aluminum. They have found that PCD tool exhibits superior results with respect to roughness, tool wear, and smoother chip under face [17]. Kumar et al. studied the performance of the TiN coating of the CBN tools during machining of the AISI4340 steel. It was observed that optimum machining performance was achieved at cutting speed of 150 m/min, feed at 0.1 mm/rev, with 40 HRC of hardness, and 1.2 mm of nose radius. The cutting force reduces due to the low frictional heat generated at the tool–chip interface. The lower cutting force and cutting zone temperature reduce at lower nose radius since the area of contact resembles the amount of cutting force and tool wear formed at the tool–chip interface [18]. Mane et al. founded that cutting force and cutting temperature of AISI 4340 steel reduce by using multi-layered coated carbide cutting tool. The feed is the most influencing cutting parameter followed by depth of cut and cutting speed with a contribution percentage of 58.03%, 26.64%, and 14.01% respectively [19]. Tiwari et al. had used cermet coating insert for turning of the AISI4340 steel. The study had been focused on the chip reduction rate, surface roughness, and material removal rate. A good quality of surface finish is achieved which lies within a range of 0.212–1.452 μm. Depth of cut is the most influential cutting parameter that affects the material removal rate and chip reduction coefficient with a percentage contribution of

90.53% and 83.14%. Surface roughness is mostly affected by the feed and depth of cut with a percentage contribution of 55.03% and 44.09%, respectively [20].

2 Methodology

2.1 Taguchi's L9 Orthogonal Array

Optimization plays a vital role for better product quality with desirable rate of production in a reasonable cost that improves the production processes for selection of appropriate machining parameters. Optimization is the process of minimizing undesired output and maximizing desired output for sustainable machining performance. While, conventional layout of cutting parameters is difficult to implement when the number of cutting variables are more. To overcome with this problem, the Taguchi method suggests a robust design layout which covers the whole parametric combination within a less number of trials. Taguchi's L_9 orthogonal array is used to conduct the experiment to reduce both time and cost of experiment.

2.2 Optimization of Turning Process by VIKOR Method

The Vlse Kriterijumska Optimizacoja I Komopromisno Resenje (VIKOR) method is a multi-criteria decision-making (MCDM) process to choose the best optimal parameters for better machining performance. By using VIKOR method, all responses are converted into a single response called VIKOR index, and optimal parametric setting is found out corresponding to lower value of VIKOR index. The procedure of VIKOR method is explained as follows [21–23]:

1. Normalize the output responses.

$$Y_{ij} = \frac{y_{ij}}{\sqrt{\sum_{i=1}^{n} y_{ij}^2}} \tag{1}$$

y_j^{min} and y_j^{max} are the minimum and maximum value of output responses, respectively. Y_{ij}= normalized value.

2. Calculate the ideal (A^*) and negative ideal (A^-) solution.

$$A^* = \{\max Y_{ij,} i = 1,2.....n\} \quad \text{for higher is better or} \{\min Y_{ij,} i = 1,2.....n\}$$
$$\text{for lower is better} \quad A^* = \{Y_1^*, Y_2^*, \ldots\} \tag{2}$$

$$A^- = \{\min Y_{ij}, i = 1,2.....n\} \text{ for higher is better or}$$
$$\{\max Y_{ij}, i = 1,2.....n\} \text{ for lower is better } A^- = \{Y_1^-, Y_2^-, ...\} \quad (3)$$

3. Calculate the utility (S_i) and regret (R_i) measures for each experimental trials.

$$S_i = \sum_{j=1}^{n} \frac{w_j (Y_j^* - Y_{ij})}{(Y_j^* - Y_j^-)} \quad (4)$$

$$R_i = \max_j \frac{w_j (Y_j^* - Y_{ij})}{(Y_j^* - Y_j^-)} \quad (5)$$

w_j = weightage such that $\sum w_j = 1$

4. Calculate the VIKOR index of the ith experimental trial.

$$Q_i = v\left(\frac{S_i - S^*}{S^- - S^*}\right) + (1-v)\left(\frac{R_i - R^*}{R^- - R^*}\right) \quad (6)$$

$i = 1, 2, 3...n,$

$$S^* = \min S_i, \quad S^- = \max S_i, \quad R^* = \min R_i, \quad R^- = \max R_i.$$

ν = weight of the maximum group utility, which is usually taken as 0.5.

3 Experimental Details

AISI 4340 hardened steel of cylindrical rod with 40 mm diameter and 250 mm long is used as work piece material for this experiment (as shown in Fig. 1). The hardness of AISI 4340 hardened steel is 52 HRC. AISI 4340 steel is difficult to machine due to excessive hardness, low specific heat, and high strain rate. Due to heat treatment,

Fig. 1 Work piece

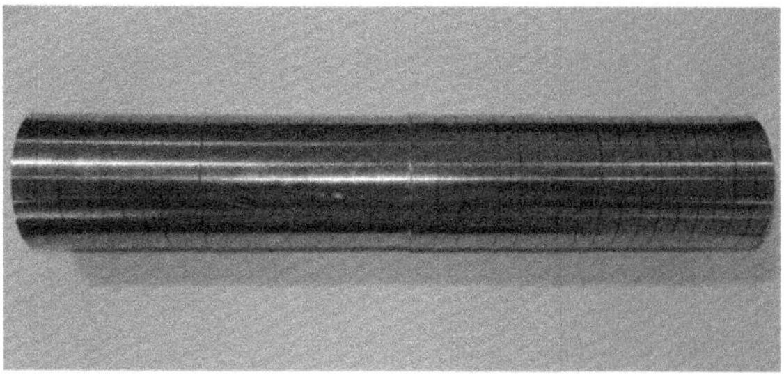

the material hardness increases. Therefore, turning of this material is called as hard turning. This steel is widely used in industries for making components like axles, gears, shafts, bearings, structures, and automobile parts. The chemical composition of the workpiece is shown in Table 1. In the present study, cubic boron nitride (CBN) cutting insert is used having specification CNMX 120408EN TA201 in dry environment. The tool insert and designated tool holder (PSLNR2525M12) are shown in Figs. 2 and 3, respectively.

Three cutting parameters taken during hard machining are cutting speed, feed rate, and depth of cut with three levels and their influence on output responses like machining forces (axial, radial, and tangential force) and surface characteristics (average surface roughness, height of the profile, average height of the profile) of the machined surface are analyzed. Taguchi's L_9 orthogonal array is used to conduct the experiment to reduce both time and cost of experiment. All the cutting forces are measured by a three component force dynamometer provided on the turret disk of the high-speed precision lathe through a custom designed turret adapter for tool holder generating sufficient rigid tooling fixture. The forces acting towards the tool are essential part of machining to analyze the machinability conditions. The surface roughness parameters of the machined surface are measured by surface roughness profile-meter (Taylor Hobson Surtronic 128). The different levels of turning parameters are given in Table 2.

Table 1 Chemical composition of AISI 4340 steel

Element	C	Si	Mn	Ni	Cr	Mo	Fe
Weight (%)	0.42	0.20	0.72	1.80	0.75	0.23	Rest

Fig. 2 CBN insert

Fig. 3 Tool holder

Table 2 Turning parameters with their levels

Parameters	Actual values	Unit
Cutting speed (v)	106, 138, 202	m/min
Feed rate (f)	0.08, 0.1, 0.16	mm/rev
Depth of cut (d)	0.2, 0.4, 0.6	mm

Table 3 Responses for turning operation

SI. No	v (m/min)	f (mm/rev)	d (mm)	F_x (N)	F_y (N)	F_z (N)	R_a (μm)	R_z (μm)	R_t (μm)
1	106	0.08	0.2	74.46	109.84	165.34	0.61	3.5	5.5
2	106	0.12	0.4	102.85	150.58	248.05	0.69	4	6
3	106	0.16	0.6	184.23	318.46	425.34	0.95	5.5	8.12
4	138	0.08	0.4	98.34	125.21	211.32	0.56	3.5	5.5
5	138	0.12	0.6	134.34	196.91	252.45	0.67	4	6.57
6	138	0.16	0.2	82.57	116.12	160.93	0.82	4.5	7.39
7	202	0.08	0.6	126.48	172.98	215.72	0.52	3.35	4.9
8	202	0.12	0.2	50.5	88.54	132.59	0.53	4	6.45
9	202	0.16	0.4	111.36	165.94	219.45	0.76	4.36	5.9

The values of each parameter during all the nine experiments are also shown in Table 3. The experiment of turning operation is performed in AISI 4340 steel by using CBN tool insert and corresponding responses like machining forces F_x, F_y, and F_z and surface roughness parameters R_a, R_t, and R_z are measured and presented in Table 3.

4 Results and Discussion

4.1 Optimization of Turning Process by VIKOR Method

Optimization is the process of minimizing undesired output and maximizing desired output for sustainable machining performance. In this turning operation, all the output parameters like F_x, F_y, F_z, R_a, R_t, and R_z are minimized.

By following the procedure of VIKOR method as described in Eqs. (1)–(6), the normalized responses, utility (S_i) measures, regret (R_i) measures, and VIKOR index (Q_i) are calculated and presented in Table 4. The results are analyzed by analysis of variance (ANOVA) and main effect plot generated by using MINITAB-17 software. The ANOVA of Q_i is given in Table 5 with $R^2 = 94.5\%$. The significance of machining parameters on the responses is determined by ANOVA. It is observed that all the three parameters are insignificant. All the three parameters p-values are more than 0.05 with 95% confidence level. The feed rate, depth of cut, and cutting speed are contributing about 57.56%, 32.28%, and 4.64%, respectively. $R^2 = 94.5\%$ shows that the model predicts the response with high accuracy. From Fig. 4, it is observed

Table 4 Normalized responses (Y_{ij}), utility (S_i) measures, regret (R_i) measures, and VIKOR index (Q_i)

Sl. No.	F_x (Y_{ij})	F_y (Y_{ij})	F_z (Y_{ij})	R_a (Y_{ij})	R_z (Y_{ij})	R_t (Y_{ij})	S_i	R_i	Q_i
1	0.2190	0.2116	0.2301	0.2937	0.2827	0.2896	0.1415	0.0349	0.0041
2	0.3026	0.2901	0.3452	0.3322	0.3231	0.3159	0.3491	0.0659	0.2416
3	0.5420	0.6136	0.5919	0.4574	0.4443	0.4276	0.9999	0.1667	0.9999
4	0.2893	0.2412	0.2941	0.2696	0.2827	0.2896	0.1892	0.0596	0.1254
5	0.3952	0.3794	0.3513	0.3226	0.3231	0.3460	0.4462	0.1045	0.4441
6	0.2429	0.2237	0.2240	0.3948	0.3635	0.3891	0.4103	0.1289	0.5159
7	0.3721	0.3333	0.3002	0.2503	0.2706	0.2580	0.2032	0.0947	0.2665
8	0.1486	0.1706	0.1845	0.2552	0.3231	0.3396	0.1344	0.0802	0.1720
9	0.3276	0.3197	0.3054	0.3659	0.3522	0.3107	0.4044	0.0930	0.3765

Table 5 ANOVA for Q_i

Source	DF	Seq SS	Adj SS	Adj MS	F	P	% contribution
Speed	2	0.03158	0.03158	0.01579	0.84	0.543	4.64
Feed	2	0.39139	0.39139	0.19569	10.43	0.087	57.56
Doc	2	0.21945	0.21945	0.10973	5.85	0.146	32.28
Error	2	0.03753	0.03753	0.01876			5.52
Total	8	0.67995					100

that Q_i increases with increase in feed rate and depth of cut, whereas it decreases with increase in cutting speed. From Table 6 and Fig. 4, it is found that the optimum levels for lower values of machining forces and surface roughness parameters are level 3 (202 m/min) for speed, level 1 (0.08 mm/rev) for feed, and level 1 (0.2 mm) for DOC, respectively.

4.2 Tool Wear

In this research, the cutting tool was made with cubic boron nitride (CBN). This tool has properties like low frictional coefficient and anti-adhering. The adhesion of steel on the cutting edge is reduced by decreasing cutting resistance which reduced formation of built-up edge. As a result, it led to maintain the sharpness of the edge with superior machining quality of the work materials. The major wear forms on the tools are the combination of flank wear and rounding of the nose. During early machining period, the flank wear was seen at the cutting edge and the outer layer of coating material was rubbed away due to abrasion between the insert and the work piece. During machining, the wear increased and material was removed from the cutting edge and then the flank wear was also increased quickly till the cutting tool had to be discarded. This is called one type of wear which is chip out wear. Figures 5 and 6 show the SEM image of tool insert before and after machining. During turning the cutting zone temperature increases, this softens and decreases the strength of

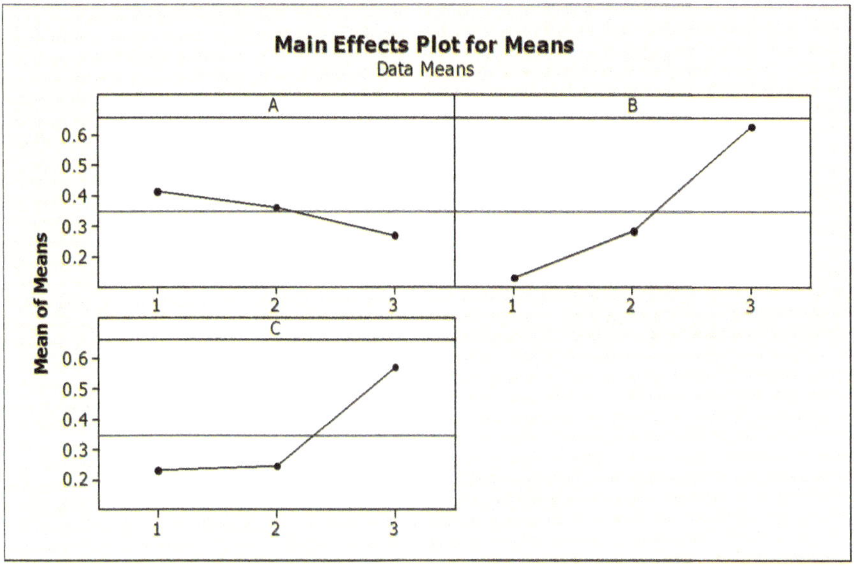

Fig. 4 Main effect plot for Q_i

Table 6 Response table for Q_i

Level	1	2	3	Delta	Rank
Speed	0.4152	0.3618	0.2717*	0.1435	3
Feed	0.1320[a]	0.2859	0.6308	0.4988	1
Doc	0.2307[a]	0.2478	0.5702	0.3395	2

[a]Optimum level

the BUE. Mainly the tool flank wear was strongly affected by the interactions between the cutting tool and work piece in the form of contact stress and cutting temperature during operation. Due to increase in cutting speed and feed rate, the rubbing action becomes faster and more heat is produced although the contact time reduces, which generates heat at flank side and softens the edge and increases the wear on the tool surface. It is also one type of tool wear which is also known as chip out from the surface of the cutting tool. Sometimes white layers were observed with hardness more than that of the bulk material which is showing in Fig. 7 (optical images of tool insert after machining). There was a reduction of iron and chromium, whereas increase in carbon and oxygen content on the white layer. The tool wear rate is increased with increase in cutting speed and the depth of white layer and hardness is reduced. This is due to increase in speed. The temperature of work piece material is reduced with increase in temperature of chip. This reduce in temperature of machined surface is due to faster rate of chip removal and deficient contact time. Therefore, less heat is conducted into work piece, whereas more heat is carried out by the chip. This study was observed during the hard turning of AISI 4340 steel with CBN insert.

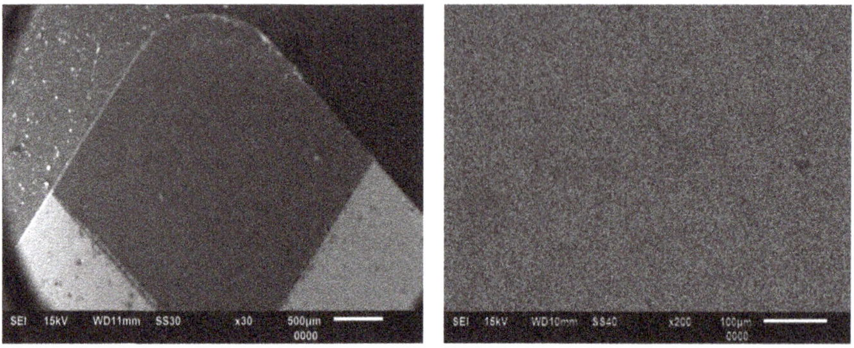

(a) (b)

Fig. 5 SEM images showing the CBN tool before machining

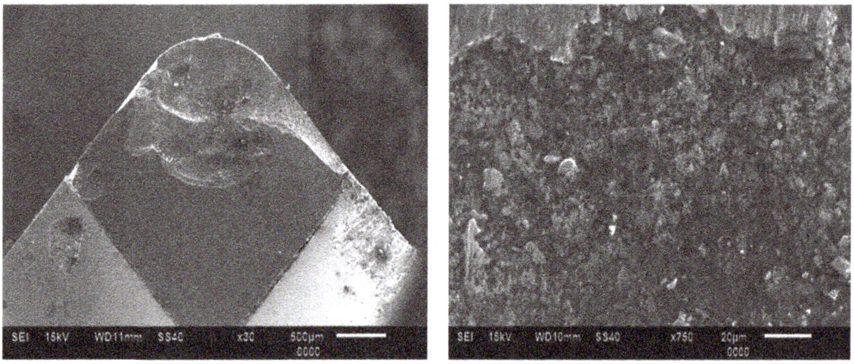

(a) (b)

Fig. 6 SEM images showing the wear of CBN tools after machining

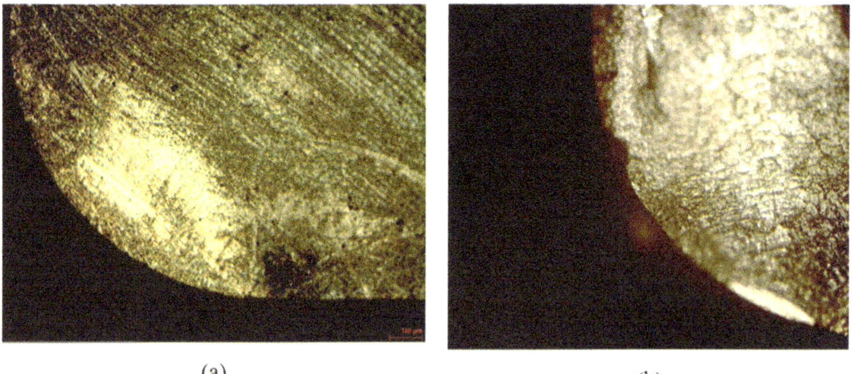

(a) (b)

Fig. 7 Optical images of tool insert after machining at (**a**) Speed = 106 mm/min, Feed = 0.16 mm/rev, Doc = 0.6 mm and (**b**) Speed = 138 mm/min, Feed = 0.0.08 mm/rev, Doc = 0.4 mm

5 Conclusion

The current study analyzed the influence of cutting speed, feed rate, and depth of cut on the performance of cubic boron nitride (CBN) tools in terms of cutting forces, surface roughness parameter while turning AISI 4340 steel. The MCDM approach like VIKOR method is used for the multi-response optimization of turning operation. The analysis of variance (ANOVA) result of VOKOR index shows that the optimal combination of low feed rate and low depth of cut with high cutting speed is advantageous for reducing machining forces and surface roughness parameters collectively. The feed rate, depth of cut, and cutting speed contribute about 57.56%, 32.28%, and 4.64%, respectively. The study concludes that hard turning variables (cutting forces, surface roughness parameters) are essential for the optimization of cutting tool design and cutting conditions (cutting speed, feed rate, depth of cut) such that product quality, productivity, and tool life are maximized. The hard turning process is being used gradually in industry to replace the costly and slow grinding process in finishing mechanical components.

References

1. Yang, W. H., & Tarng, Y. S. (1998). Design optimization of cutting parameters for turning operations based on the Taguchi method. *Journal of Materials Processing Technology, 84*, 122–129.
2. Nalbant, M., Gokkaya, H., & Sur, G. (2007). Application of Taguchi method in the optimization of cutting parameters for surface roughness in turning. *Materials and Design, 28*, 1379–1385.
3. Rathod, K. B., & Lalwani, D. I. (2017). Experimental investigation of flank wear and surface roughness during hard turning of AISI H11 steel with CBN tools. *Indian Journal of Engineering & Materials Sciences, 24*, 171–181.
4. Das, A., Mukhopadhyay, A., Patel, S. K., & Biswal, B. B. (2016). Comparative assessment on machinability aspects of AISI 4340 alloy steel using uncoated carbide and coated cermet inserts during hard turning. *Arabian Journal for Science and Engineering, 41*, 4531–4552.
5. Asilturk, I., & Akkus, H. (2011). Determining the effect of cutting parameters on surface roughness in hard turning using the Taguchi method. *Measurement, 44*, 1697–1704.
6. Stakhniv, N. E., & Devin, L. N. (2012). The study of the influence of wear of the cutting tool with an insert of cBN-based composite material on the tool vibrations in finish turning of hardened steels. *Journal of Superhard Materials, 34*(3), 193–199.
7. Gangopadhyay, S., Acharya, R., Chattopadhyay, A. K., & Sargade, V. G. (2010). Effect of cutting speed and surface chemistry of cutting tools on the formation of BUL or BUE and surface quality of the generated surface in dry turning of AA6005 aluminium alloy. *Machining Science and Technology, 14*, 208–223.
8. Bhushan, R. K., Kumar, S., & Das, S. (2010). Effect of machining parameters on surface roughness and tool wear for 7075 Al alloy SiC composite. *International Journal of Advanced Manufacturing Technology, 50*, 459–469.
9. Che-Haron, C. H. (2001). Tool life and surface integrity in turning titanium alloy. *Journal of Materials Processing Technology, 118*, 231–237.
10. Nordgren, A., Samani, B. Z., & Saoubi, R. (2014). Experimental study and modelling of plastic deformation of cemented carbide tools in turning. *Procedia CIRP, 14*, 599–604.

11. Ramesh, S., Karunamoorthy, L., & Palanikumar, K. (2012). Measurement and analysis of surface roughness in turning of aerospace titanium alloy (gr5). *Measurement, 45*, 1266–1276.
12. Jagadesh, T., & Samuel, G. L. (2014). Investigations into cutting forces and surface roughness in micro turning of titanium alloy using coated carbide tool. *Procedia Materials Science, 5*, 2450–2457.
13. Upadhyay, V., Jain, P. K., & Mehta, N. K. (2013). In-process prediction of surface roughness in turning of Ti–6Al–4V alloy using cutting parameters and vibration signals. *Measurement, 46*, 154–160.
14. Khan, A., & Maity, K. (2017). Parametric modelling of multiple quality characteristics in turning of CP titanium grade-2 with cryo-treated inserts. *International Journal of Materials and Product Technology, 54*(4), 306–331.
15. Khan, A., & Maity, K. (2018). Influence of cutting speed and cooling method on the machinability of commercially pure titanium (CP-Ti) grade II. *Journal of Manufacturing Processes, 31*, 650–661.
16. Parida, A. K., & Maity, K. (2018). Comparison the machinability of Inconel 718, Inconel 625 and Monel 400 in hot turning operation. *Engineering Science and Technology: An International Journal, 21*, 364–370.
17. Pattnaik, S. K., Bhoi, N. K., Padhi, S., & Sarangi, S. K. (2018). Dry machining of aluminum for proper selection of cutting tool: Tool performance and tool wear. *International Journal of Advanced Manufacturing Technology, 98*, 55–65.
18. Kumar, S., Singh, D., & Kalsi, N. S. (2020). Performance evaluation of TiN-coated CBN tools during turning of variable hardened AISI 4340 steel. In *Advanced engineering optimization through intelligent techniques* (pp. 847–857). Springer.
19. Mane, S., Mishra, A., & Kannawar, V. (2020). Optimization of cutting parameters in dry turning of AISI 4340 hardened alloy steel with multilayered coated carbide tool. In *Proceedings of international conference on intelligent manufacturing and automation* (pp. 99–105). Springer.
20. Tiwari, P. K., Kumar, R., Sahoo, A. K., Panda, A., Das, D., & Roy, S. (2020). Performance evaluation of coated cermet insert in hard turning. *Materials Today: Proceedings, 26*, 1941–1947.
21. Sahu, A. K., Mahapatra, S. S., & Chatterjee, S. (2018). Optimization of electro-discharge coating process using harmony search. *Materials Today: Proceedings, 5*, 12673–12680.
22. Sahu, A. K., & Mahapatra, S. S. (2019). Optimization of surface roughness parameters by different multi-response optimization techniques during electro-discharge machining of titanium alloy. In K. Kumar, N. Kumari, & J. P. Davim (Eds.), *Non-conventional machining in modern manufacturing systems* (pp. 82–108). Springer.
23. Khan, A., & Maity, K. A. (2016). Novel MCDM approach for simultaneous optimization of some correlated machining parameters in turning of CP-titanium grade 2. *International Journal of Engineering Research in Africa, 22*, 94–111.

An Experimental Study on Major Process Parameters Effecting the Type of Burrs in Drilling Operation for Mild Steel ASTM A-36

Anas Islam and Vijay Kumar Dwivedi

1 Introduction

Drilling is among the most basic and complicated production methods. Burr is the most frequently occurring drilling issues. Such burrs intervene in parts formation, causing uncertainty and disassociation [1–3]. The ridges pose issues with durability or loss of efficiency of particular sections. The system burrs result in various unacceptable factors in operation, including inappropriate interaction with current transport modules and unsuitable seating of pairing surfaces [4, 5]. Burrs are detrimental to the leading edge and the groove tears as they are machined. In terms of consistency and efficiency, the creation of exit burrs on component borders while boiling has certain negative aspects. There are hardly any methods required for deburring when the escape burr is shaped inside a cavity. Rather specific equipment can be added, thus improving the efficiency of deburring. The deburring method is normally performed manually due to software issues, takes additional time, and can harm the edges, which result in the dismissal of the product [6]. The end finish of precision parts will also cost a ton. The factors influencing the development of burrs at the outlet of the holes in boiling are therefore necessary to be understood to minimize burr size during growth. This work aims primarily at carrying out this significant topic over the past few years. The chapter will address burr forming in different processes of optimization of the perforation by means of twist boxes determined by the cutting conditions and the drill geometry. Boiling optimization to determine optimum value for a defined drilling diameter that simultaneously decreases burr scale, that is, burr height and burr density, for tool geometry, feed, inclination of the tip or lip clearance angle. Milling, cutting, grinding, gravure, or turning burr forming was one of the key challenges for the correct mining, material manufacturing, and

A. Islam (✉) · V. K. Dwivedi
Department of Mechanical Engineering, IET, GLA University, Mathura, India

C. Prakash et al. (eds.), *Additive, Subtractive, and Hybrid Technologies*,
Mechanical Engineering Series, https://doi.org/10.1007/978-3-030-99569-0_2

distribution fields. The consistency and accuracy of goods are compromised by several various forms of burr, and further deburrings are induced. Yet deburring accounts for a large portion of manufacturing costs. It is also really necessary to mitigate this. In this chapter, we discuss the development of burrs during boiling at the entrance and exit of the holes in the production level. Precision components are needed to manufacture all surfaces and measurements as well as diameter with very close tolerances when manufactured in contemporary manufacturing. Good quality of goods can be manufactured with minimum engineering of processing costs in compliance with design specifications. In order to accomplish these goals, it is important to recognize the production cycle and reduce its parameters. The typical manufacturing methods include machining, metal shaping, molding, and casting injections. Drilling is one of the most common and dynamic methods in several various forms of machining. This is commonly employed in other fields, including the aerospace and automobile sectors. The projection of the substance, known as "Burr," acts as one of the main size and dimension errors during drilling. Much of the method of machining creates burrs. The burr is a coating that is extracted during machining from the boundaries or surfaces of a workpiece. Burr is a deformed plastic substance created on the edge of the component during cutting or shearing. In both the cases of boiling, these burrs were noticed when the chisel edge came in touch with the workpiece, directly at the beginning of the cutting. Once the drill is inserted into the workpiece, the burrs are shaped and dispersed in a circumventing path.

1.1 Formation of Burrs

Indubitably extremely nuanced is burr work. A major concern is the lack in knowledge of burrs and the process of burr forming for aviation and auto industries. The procedures also create rim faults of the generated component in the production system. The type of jutting rough product around the sides of the component called as burrs will take these surface deficiencies. Gillespie and Blotter divided the system of development burrs into four separate kinds: shrimp, rolled over, and break burrs as seen in Fig. 1, and cut off burrs. The ranking is focused on their development process. The fish burr or stress burr is a result of a propensity of tissue to bulge when squeezed at the edges. The substance is squeezed in the context of Poisson burr till the product becomes fatally deformed. The most frequent form of burr is a burr that is bent instead of screwed, particularly by a chip. At the end of a break is the large burr. The burr or burr is induced by removing material loosely from work material instead of moving. The disconnected burr arises from the removal of the raw material from the component, before the disconnection is done. In the style of entering burrs, fish is mostly included, whereas the style of rust and rust is primarily for boiling exit burrs. Gillespie and Blotter have described three essential processes involved in burr creation: longitudinal substrate deformation, chip cracking, and chip breaking. Nakayama and Aral [7] also suggested an alternate yet identical classification model.

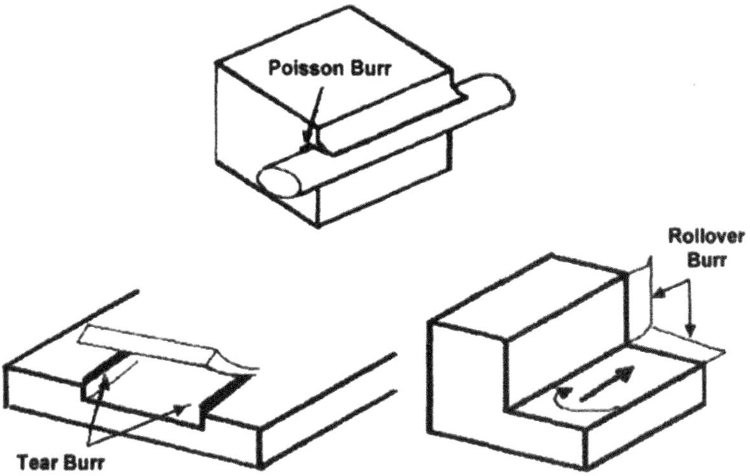

Fig. 1 Basic types of burrs

The burr structure and composition, which are divided into three parts: creation, production, and ultimate burr creation, were also developed by Ko [8] and Dornfeld. The technique is focused on evaluation while cutting process in machining experiments on strain hardening. 3D burr forming in angular clipping is studied by Hashimura et al. [9]. The cut studies have been conducted using a micro tool on the "scan electron microscope" (SEM). Immediate negative shear inclination characterizes the start of burr forming as the instrument gets to the edge of work material [10]. The end state on the bottom of the negative shear plane serves like a rubber hinge throughout the production period. The adverse shear zone spins at this level as the device goes on [11, 12]. Eventually, when the instrument reaches the end of the work material, a burr develops due to the rising pressure along detrimental shear line. On the outer edge of machined surface outer burr and then on the inside burr on the surface of the workpiece is found. It was discovered that angular cutting outlet burr is lesser than for the orthogonal, whereas the angular cut side burr was bigger than orthogonal burr [13–16].

1.2 Parameters that Influence the Drilling of Burr

Drilling burr forming is a dynamic theoretical method since it is a three-dimensional method and is inherently influenced by several factors such as the structure of the drilling, material properties, and the conditions of the field. The fundamental information of burr formation process is acquired by experimental data recovery and study. Stein [1] analyzed, incorporating fractional factorial method of experimental studies, the forming of burr for the specific drilling of miniature holes in stainless steel 304L content. The effects on burr height, thickness, and shape were recorded

Table 1 Parameters affecting drilling burr formation

S.No	Category	Parameters
1	Process conditions	Cutting speed, feed, use of coolant
2	Drill geometry	Point geometry, point angle, lip clearance angle, helix angle
3	Material properties	Ductility, hardness, tensile toughness, strain hardening characteristics
4	Others	Temperature dependence properties

of feed pace, traverse speed, staring intently, and tool content. The researchers of Gillespie [2] were one of the first to examine burr shape in drilling at academic point. In a number of test environments, he examined the impact of process parameters, drill geometry, and material properties on burr shape. The research of Gillespie [3] on titanium alloys discussed the issues of hole strength and underlined the effect on burr size of drilling wear property. In this analysis, no effect has been detected. Ko and Dornfeld [4, 5] suggested a model of quantitative burr forming for orthogonal cutting ductile products. In a "scanning electron microscope (SEM)," machining experiments were carried out to determine the effect of the machining parameters on burr thickness [6]. In addition, equations have been developed for calculating burr height and burr thickness, depending upon cuts, chip morphology, and work material, such that the shear angle and the contact length of the tool-chip are calculated. Gillespie and Blotter [8] established basic research frameworks for fish and rolled over burrs that could determine burr thickness for particular circumstances with certain results. Furthermore, the process of burr forming was regulated by the basic factors such as thrust, uncut chip size, comparative energy to twist, shave chips, etc. Some of Gillespie's [12] trials have been carried out with hand feeding drills (unknown and unchecked feeding rates). The effect of feed levels is often mixed with other observed parameters. In the drill of solid carbide and high velocity cobalt boilers Ti-6AI-4V titanium alloy tubes, Dornfeld et al. [13] also examined the impact from device design and plant state on burr formation. The circumstances of the cutting had no impact on the size of the burr, and the layout of the drilling such as helix, dividing point versus helicopter, lip relief angle, and point angle had a considerable influence on burr thickness and burr height. In a previous work by Sofronas et al., several aspects have affected drilling burr shape. Among these variables that are more important are workpiece content, drill configuration, and process conditions, and most of the analyses have centered mostly on the impacts of the drilling burr shape. Table 1 displays the control criteria that lead to burr forming.

1.3 Drilling Burr Learning Mechanism

Drilling is perhaps a rotational movement chips form operation. The tool's feed motions are just in the rotary axis orientation. The most frequently employed and readily accessible drill configuration is the 2-flute triangular spike drill using in

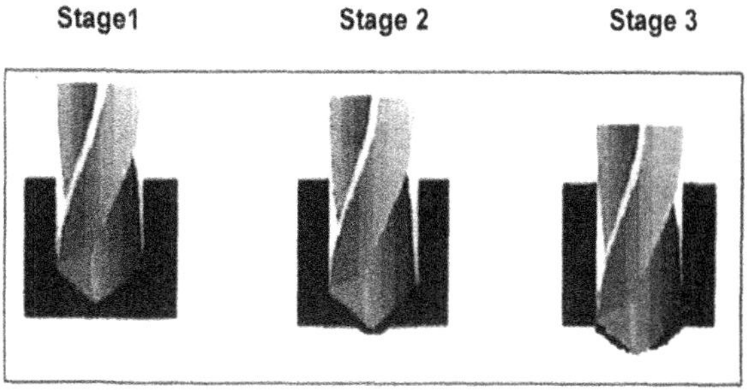

Fig. 2 Three stages of exit burr formation in drilling

spherical hole creation. Drilling burr development is a very complex surface displacement action [10, 17–19]. The burr forming method involves material characteristics, component structure, surface preparation, machine configuration, machine direction, and parameters of cutting [20–23]. There are many considerations implicated in burr forming. Two forms of burrs, a tiny entry burr and a much larger exit burr, are produced during boiling [24–28]. The entry burr is built in the shape of a tiny wedge across the boiling hole, but the escape burr emerges at the opposite side as the boiler pierces the job component through unaltered volumes. The boiling out cycle may be categorized in three steps as seen in Fig. 2.

Phase 1: Bulge rises on the layer on the exit. The edge of the chisel side of the drill produce the plastic deformation at the bottom of the workpiece as the boiling reaches the workpiece's exit [29–34].

Phase 2: Under point content stretches to full duration. At the edge of the work material, a bulge forms. The remainder of the material is also solid enough to get to the corners of the device to endure the thrust energy [35–39]. So there is no permanent deformation in this zone, and hence, the usual phase of cutting begins.

Phase 3: Drill point appears from the exit surface of the workpiece. When it approaches the full flexion, the layer under the blade surface continues to break as the boil eventually breaks in, the remainder is drawn out or become buried [32, 40–44].

1.4 Characterization of Burr

The burrs are typically identified by the length and width of the burr. As seen in Fig. 3, this procedure is used to determine burr height and thickness.

The length of burr is identical to the feeding direction of hammer, from unmade workpiece output layer to the end of burr at every level along the diameter of burr. The thickness of a burr is the breadth of the burr root usually measured to the drill's feed axis [45–51]. The origin of the burr is in the section of burr, from the outer regions of the cavity to the end location of the permanent deformation at every level along the radius of the tube [8, 52]. The height of the burr is used to assess the period required to deburr. The thickness of burr is seen as the biggest

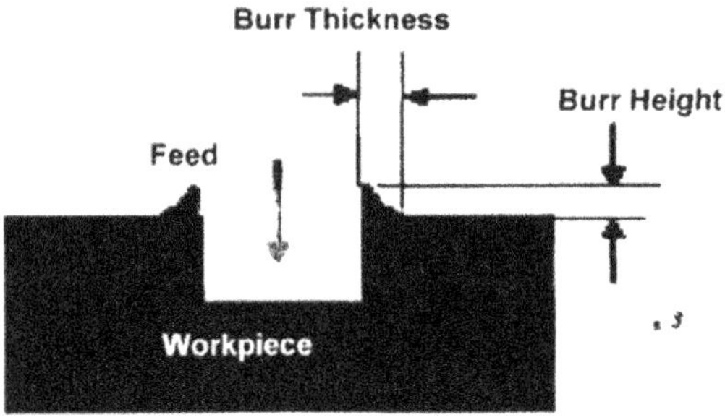

Fig. 3 Burr characterization

obstacle to deburring. With practical manufacturing processes and environments, Dornfeld [53] extended the research paradigm of burr forming. Their research concentrated primarily on the creation of burrs and breakouts of bottom. Nevertheless, there are no analytically or empirically valid equations to forecast and monitor burr forming in oblique cutting processes. Sofronas and Taraman [54] have carried out theoretical and experimental work and have apparently lowered the exit burr density by growing the helix angular and the lip clearance angles and decreasing the feed angle. Shikata et al. [55] have recorded related finding patterns in carbon steel sheet drilling under specific experimental conditions and suggested the framework for classifying burr scale. In order to collect information on the cutting force, traction, and hole size, determined by the design of the burrs shaped on the workpiece, five drills each with a specific point shape were used. Sofronas [56] experimentally examined the effect on the scale of exit burrs when drilling steel of working material properties as shear strength and hardness. The exit burr has demonstrated a marginal impact on burr size with increasing strength and device stiffness. The volume of the burr has been seen to rise, as the drill diameter has increased from 0.2 to 2.5 mm as the number of burrs decreases with diameters of boilers smaller than 0.2 mm in size. In the first example, the action was attributed to a smoother curve with a decrease in the boiling diameter. For the above scenario, the trigger was not apparent. The method of burr forming was also researched by Sugawara and Inagaki [57, 58] using model experiments. The effect of the boom types and working practices on the development of burrs have been studied. The sluggishness on the corners induced the creation of a wider burr such that the cutting capability increased [59]. The burr height was found to differ based on the grain orientation in polycrystalline copper boiling. As the radius of the tool employed was in the order of grain thickness, the cutting mechanics that exist on the edge of a blade differs in most situations with the device going from grain to grain. One substrate may create a ductile mode like the cutting mode in one grain and brittle like the cut in another, meaning that favorable and unfavorable cutting directions occur according to crystallographic inclination for a good surface and edge conditions.

Takazawa [60] has studied different strategies to study the impact of component content on boiling burr shape. The boiling burrs of a dubbed drill on cutting lips is said to be smaller than the burrs that standard drills make. The decline in burr size was because the axial force was decreased, and the chip flow into a hole became gradually smooth. The investigation was done by Kim [61] to explore the development of a boiling burr on titanium alloy Ti-6 A1-4V which is more often used in the aviation industry because of its high specific power. The impacts of differences in feed rate and cutting speed on the forming in drilling burrs were found utilizing two separate kinds of carbide borers. There was no need of coolant. Mass height and burr thickness measurements did not provide very valuable results, as small burrs were shaped in all circumstances in fairly homogeneous circumstances. In the drilling of low alloy steel, Kim and Dornfeld [62] established an experimental model for burr forming. The suggested process for burr forming was based on energy efficiency and the principle of metal cuts. The results on burr forming were evaluated on the basis of the framework, some essential parameters. In the Drilling Burr Control Chart (DBCC), Kim [63] has developed a prediction algorithm for the estimation of boiling burr and data updating method. The probable forecast offers a more workable method for monitoring the creation of burrs in mass processing, and the enhancement technique reveals how new data collected can be utilized through subsequent drilling. Dependent on the method parameters for a reliable mixture of the substance and the configuration of the drills, this graph indicates the different distributors of the burr types. This research has applied a Bayesian parametric simulation method. The density function of the beta frequency defined the density of the likelihood of the creation of a given category of burr. This is useful for modeling probability behavior, which requires the creation of such response variable which are expected to fall inside the period (0, 1).

Heisel et al. [64] proposed a procedure for deciding the burr measurements in short hole boiling, taking into consideration the determinants that affect the shape of the burr. The resultant heat, forces, and structure of inserts are such parameters. The approach is based on observational cutting experiments and takes the association between various burr parameters and the machining conditions, including cutting speed, feed, and design of the machine into account. The influence of dry machining on burr size was studied by Shefelbine and Dornfeld [65]. They indicated that drying without coolant could be helpful because coolant usage costs were minimized, and likely adverse impacts on workers' safety and the atmosphere decreased. However, owing to high temperatures, many issues with dry machining exist. This has been shown that burrs have been produced at higher temperatures because of the improved ductility. The chamber depth may be lowered mostly on edge of a section, the chosen deburring method must, in order to guarantee a full borrel elimination, be able to reach a minimum chamber depth [66, 67]. The thickness of the burr therefore influences the energy and time required to deburr [68, 69]. Burr size is a significant means of quantifying burr attributes, and therefore work has already been carried out on burr size estimation owing to the need for automated deburring [70–73].

2 Boiling Burr Types

The direction of the fracture depends on the orientation of the original fracture during burr forming. Burr form is important as it depends on the size of the burr, and therefore on the cost of deburring [74–78]. The details were clarified by traditional burr forms shown while drilling in different materials and the forming mechanism.

2.1 *Drill Cap Standard Burr*

The word "uniform" applies to a broad degree of continuity across the whole diameter of burr and to a disparity in burr width and width as seen in Fig. 4.

During the ultimate stage of the boiling process, a drilling cap is shaped which may be often added to a workpiece or is removed at the end. The substance beneath the chisel edge starts to distort as it reaches the workpiece [20, 79–83]. Depression occurs primarily due to the thrust force while perforation, the difference between the escape region and the stage. When the drill moves along the field of plastic deformation stretches from the core to rim of the boiler. The deformation of the substance is known to be exceedingly rare [84–90]. The ultimate stage is to establish an original crack at the ends of the tool tip. The residual content is moved and bent before the hammer, creating a "uniform burr" with boiling tip, as shown in Fig. 4. AISI 304L grade stainless steel and AISI4118 low iron steel products reveal standard burrs with boiling cap forms [91–99].

2.2 *Cap-Less Dress Burr*

The standardized burr with no boiling cap is seen in Fig. 5. Burr shapes such kinds in fairly delicate material. When the path to the exiting layer starts, demonstrated in prior situations bulge forms [100, 101]. Yet, because of the lack of ductility of the steel, the

Fig. 4 Uniform burr with drill cap

Fig. 5 Uniform burr without cap

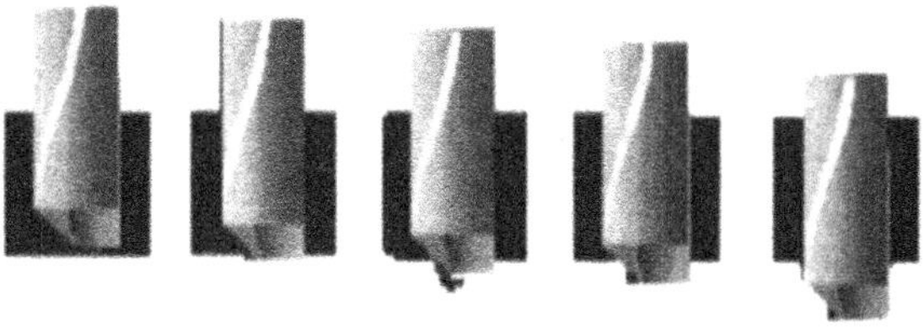

Fig. 6 Burr formation kinematics (uniform burr without cap)

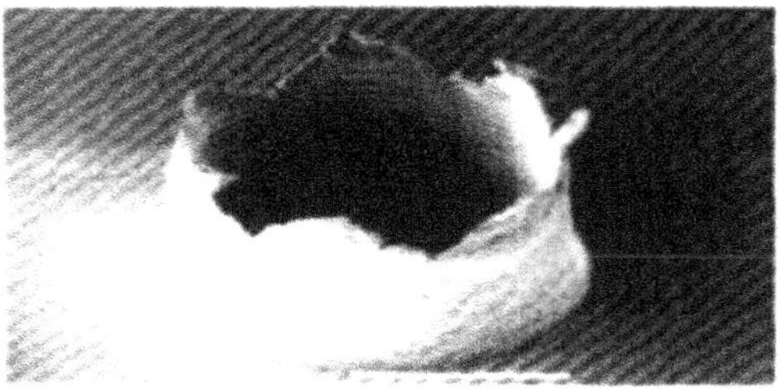

Fig. 7 Crown burr

chisel tip permeates the workpiece before inducing any permanent deformation to the object. The cutting lips take effect constantly, while the process proceeds to advance, before relatively little residual content eventually is the last stage burr as seen in Fig. 6. This method shapes burrs that are very low in height and density [102–107] and have a really uniform outline all along the circumference. In applications where the removal of drill caps is challenging, this form of burr is favored [108, 109].

2.3 Burr Crown

A crowned burr implies the burr's length and width relative to that of the normal "uniform" burr which is very broad, and the diameter of the cavity is very much distorted [110, 111], as shown in Fig. 7.

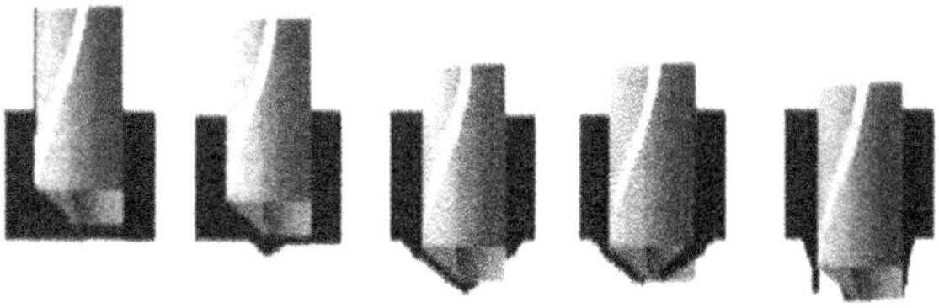

Fig. 8 Burr formation kinematics (crown burr)

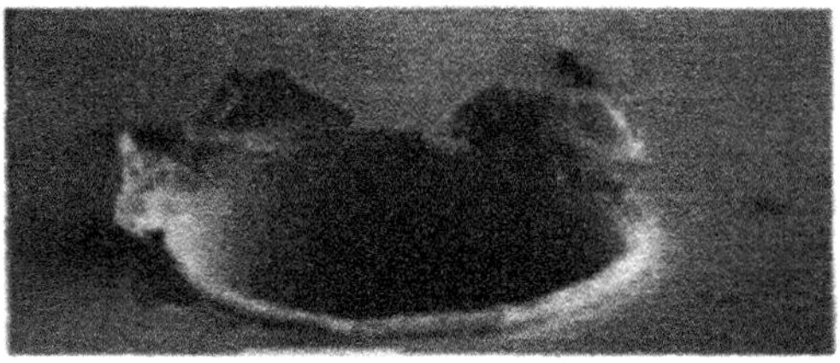

Fig. 9 Transient burr

A deformation is induced sooner by a greater thrust power. Throughout the central area of the escape surface, the thicker substance coating under the boiler undergoes plastic deformation and a greater overall pressure [112–116]. The fracture happens if this cumulative strain surpasses the fracture pressure of the element. Nevertheless, an original crack is most probable to develop in the central part of the escape field around the edge of chisel and manifest in a crown burr [117–122]. If the outer cutting edge of the exercise is considerably healed or built up, effective cutting is not required, and the substance under the exercise is more forced forward than sliced. The risk for initial cracks in the central area and the formation of the crust are enhanced, as shown in Fig. 8. In this circumstance. This method is not followed by the creation of a perforating cap [123–126], and the final burr scale is much broader than that of a flat burr. A curved burr is to be prevented from the point of view of dismantling [127].

2.4 *Temporary or Transient Burr*

In AISI 4118 lower iron steel, the persistent burr can be seen as illustrated in Fig. 9.

The temporal burr is a sort of burr developed between a uniform burr and a crown burr in the transition era. The early fracturing is identical to standard burred form at the finish of the sharp edges providing a wider rectangular region. When the stone begins to move, the pressure on the chisel's edge surpasses the material's fracturing tension [119, 124, 128–131]. The characteristics of this phenomenon are the components, the higher relative coefficient of stress strength, and the ductility. For the AISI 304L and AISI 4118 materials, three burr features are described as: small uniform burr-type I, broad uniform burr-type II, and crown burr-type III, while transient burr is composed of low alloy steel AISI 4118 only [132–135]. In boiling Ti-6AI-4V titanium alloys, Dornfeld et al. performed the experiment on burr

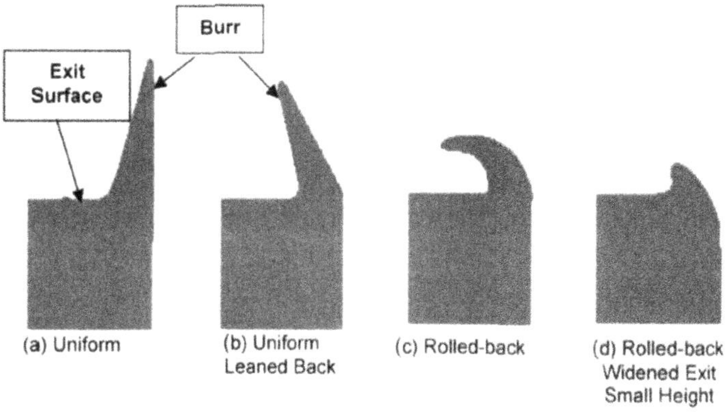

Fig. 10 Burr types formed in titanium alloy

formation. Burrs are described in the jacket, leaning back, rolled back, and rolled back shape of many classes [136–141], including the re-widening production as seen in Fig. 10. The key factor that influences the final burr formed is the production of heat because of friction between the drills and the working component during boiling.

The standard burr shapes shaped under most of the experimental cutting conditions were rolled back burrs, and the rolling back amount appeared proportional to feed rate and spinal length [142–148]. In this experiment, neither crown burr nor petal burr shaped at high feed rate in steel was produced [149]. Thermal influence induced by friction oil, because of the poor thermal conductive properties, and use of refrigerant, is assumed to have affected the forms of burrs. While boiling titanium alloy with coolant, a ring form burr was found [150–153]. The burrs produced by boiling are graded according to the position of the crack into three separate forms. Form A is the burr of the composite that is very low or negatively chamfered with a shield [154–160], shaped in the same way with a brittle material. The burr of type C is produced as the breakage starts from the middle of the cavity. "Type 1" is a borer of a little height with an uncut portion of a hole in the bottom forming as a conic hat [161–163]. "Type 2" is a larger borer shaped like a petal or crown. Experiments showed clearly that the highest temperature at the heart or at the center of the hole was observed and that the excess borers in drilling aluminum alloy are devised into two types by their formation process or shape. "Type 2" starts to evolve when the temperature of the workpiece reaches about 250 °C at the exit burr forming region [118, 129, 164–168].

3 Burrs in Drilling-Associated Issues

Burrs are among the most difficult barriers to high efficiency and automation and hence to consistency. Those burrs pose many issues with component reliability as they mess with element mounting and may pose jamming and misalignment

[169–172]. Such burrs create dimensional errors. Burrs often present durability issues which trigger machining sequence which assembly issues when the component is being employed [173]. In fact, burrs developed on modules have created other undesirable features such as inappropriate interaction with current transport components and excessive sitting of the mattress surfaces [133, 174–178]. During workmanship, butts are injurious as they strike the edge and induce loss of the groove. In effect, this wear of groove speeds up the burr group. Burrs may trigger electrical component short circuits, reduce device fatigue existence, and serve as a starting point for crack. Crossed holes function as conduits for lubricating and cooling fluids in many systems. Burrs may create vital passage blockage and friction in liquid or gas movement into ducts that could contribute to serious problems during operation [179–186]. Leakage can result in a low-quality edge in hydro-pneumatic systems. In cases with comparatively moving parts, the scratching and tear induced by burrs not only decreases the appearance of the edge but also creates noise and vibration. Therefore, burr debris will affect the moving parts severely. Throughout thermal therapy, a crack in the pieces may result in an increase in tensile tension. An attached burr could hinder file assembly in the data storage industries or the modified burr might eventually hinder file function or trigger disk collapse. In fact, burrs can pose a safety danger for the employees, because they are typically hard and may not only harm the employees but often reduce the product functions. Throughout the following deburring method, however, burrs must be extracted, so that the component meets the tolerances defined [72, 187–192]. For certain boiling systems, all entry and exit surfaces produce a burr. The production burr is far larger than the burr produced during the drilling process. The exit burr is poor for coins, precision of completion, ruggedness of the soil, and damage of the device. Moreover, a burr is an obstacle to a productive operation. In order to successfully eliminate or avoid burrs, burrs must be correctly counted [193–196]. Measuring burrs correctly is quite challenging, since burrs created by machining are irregular and extremely sharp. The right deburring approach or procedure will be prescribed if the burr structure is calculated correctly.

4 Drilling Burr Replacement Techniques

The process and technologies on deburring have been recommended to achieve high performance, low machining cost, and high machining accuracy. Depends on work content, positioning of burrs, burr at size and necessary tolerances etc., selection of a capable and effective deburring method. In fact, some of these methods are very effective, but most of them require advanced equipment and therefore some of the drawbacks of these methods, as is defined in the literature. In addition, they are used to control the structural frameworks of the device, such as mechanical, electrochemical, abrasive jet machining, ultrasonic fracturing, and laser deburring. Disposition defects on the workpiece may be induced by the burr reduction processes. Tumbles or vibratory finishing extracts materials instead of the restricted edges of the component from all surfaces of the product. The processes of tumbling and abrasive jet machining have geometric design limits that preclude them from deburring the inner sections of the component [197]. Any methods of electro-chemical removal of unwanted residuals need post-processing. Present innovation in deburring systems thus improves with both sophistication and precision. At certain times, deburring procedures such as micro-making or cross-sectioning hole boiling obviously cannot be carried out. The burr is hard to extract and the removal of the burr will harm the piece in the micro machining process. In fact, it is not feasible to conveniently implement traditional deburring for micro-burrs. Burr preparation in intersecting boils is particularly important in the automotive industry for a number of reasons. Holes that are intersected may be contained in shafts as holes in the oil inlet [82]. The burr cannot be allowed here and is because of its limited accessibility impossible to eliminate. A mechanized orbital burring is valid only for the deburring of intersections of the cross-drilled hole. Burr removal is thus always a big obstacle to factory automation and must therefore be removed from the edges of the component.

The burring of small holes, owing to limited connectivity and close tolerances, is especially challenging. There are no devices required for burring where the escape burr is shaped within a cavity. In this situation, very specific equipment will often be utilized such that the risks of burning are minimized. In the construction and production preparation of precision parts, finishing and deburring operations are sometimes ignored [198]. Such finishing operations give rise to several dimensional inconsistencies, which typically occur after a sequence of processes in the final phases of output, which add importance to the precision component. If deburring of a correct part even in the final stages of development is assumed, there is a strong probability of loss due to failure in selecting, preparing or executing the edge-finish procedure. The experiments found that the timer to deburr a part decreases exponentially with an rise in burr thickness in manual deburring of the pieces. This is calculated to contribute to as much as 30% of the expense of finished products, as well as the period spent deburring for automated machining account, for 5–10% of the overall machining time for particular components. In addition, dismantling can lead to dimensional errors, harm the surface finish, and trigger the residual stress. The technology of edge finishing is also not fast. The process in the possession of an operator relies strongly on the ability and is prone to a significant variation in efficiency.

5 Core Remarks/Discussion

The following main issues have been established in the sense of the aforementioned discussions.

- Despite recent breakthroughs in burr minimization techniques, new products and deburring system development require continuous improvement in deburring processes. Manual deburring was suggested as a replacement for some automatic

approaches. Although conventional deburring methodologies are extremely insecure and non-ergonomic, the manufacturing sectors still implement them. The user, subjected to noise, waste, dirt, and vibrations, involves continual care. Such factors make deburring mechanisms risky and complicated, which are very cost-effective because they can destroy a workpiece in their final step. Today, the development methods have to be low-cost-efficient in the factory floor among others. The burr may be used as a method to test the cutting efficiency during the removal of the material. A burr is permitted to exist in most aerospace drilling methods if it is below a certain height. The entry or access furnaces are generally not of significant concern on manufacturing floor because they are generally much smaller than the regular exit burr.
- Boiling burrs may have varying shapes and sizes based on a variety of criteria. A smaller burr will typically be favored, as it requires fewer time and costs to debur. In other uses, however, a specific burr type might be favored. The easiest approach to extract burr is to cover the substrate with a less ductile substance in drilling. But, in extremely restricted instances, this is true only. Total avoidance of burr forming is almost impossible when machining ductile products. Therefore, learning how to reduce the burr size is required.

6 Context for Analysis

In light of the findings from the previous discussions, two strategies to solving burr problems in the boiling cycle were established. Those are certain.

Approach 1: It is an indirect approach aimed at decreasing the intensity of burring operation, without compromising the component accuracy. Furthermore, due to the burr forming system, its strategy involves thorough selection of a deburring processes. Deburring costs are determined by burr thickness.

Approach 2: It is a direct approach, which attempts to minimize burr size in production by correctly managing process variables that would save deburring costs and time. A significant amount of work has been documented using an indirect approach in relation to boiling burr problem. Both of these study findings rely on various deburring methods. On the other hand, no systematic research on the reduction of burr size during production process boiling with a straightforward comparison was published. The core field of this research therefore:

- Identifies the parameters of the burr cycle that influence the size of burr and thickness of burr.
- Linkage between the size of the burr and the parameters involved is defined.
- To automate the drilling cycle and evaluate the optimum device parameter configuration that decreases burr thickness.
- A detailed way to figure out how the burr size and process variables communicate.
- An important method for optimizing the process parameters to reduce burr thickness.

7 Strategies to Raising Drilling Burrs

There have been several laboratory experiments and many theoretical efforts to eliminate burrs. This section has outlined the state of the art in boiling burr minimization and some of the contributions of authors to this subject. Three main methods have been suggested based on the literature review on the creation of techniques to reduce burrs in boiling. The first method includes refining process parameters such that burr forming is minimized and boiling structure optimized to minimize burr thickness.

7.1 Method Optimization to Reduce Burrs

In keeping with the cutting edge and style and course of the burr construction, Naqayama and Arai [17] classified the machining burrs. Many machining burrs may be classified correctly by integrating these two naming schemes. By reducing the undeformed chip thickness and shear stress of the sprockets and raise the incl. tilt, the scale of the sideward device is reduced [19]. With the fracturing at the root of the piece, the aspect of the forward burr is reduced [20]. In particular, Pande and Relekar [66] discovered the existence of burrs by adjusting the drill diameter, feed rate, hole duration to drill depth ratio, and the BHN work material with regard to the burr height and thickness at entry or exit from the troughs during drillings. The boiling diameter of 8–10 mm has also been reported to result in minimum burr exit height. Additionally, a mounting device for continuous feed changes during drilling has been designed and developed. The difference in workpiece structure on burrs produced during boiling was examined by Sugawara and Inagakis [67]. Workpieces of different structures were machined with boilers of several sizes that would reduce the burr by increasing the grain size of the workpiece. This influence, when the diameter of the boiler is very low and its edge is dull or dirty, is especially significant. Hewson [68] has studied the connections between the exit burrs, fluid cutting, back plate material support, and device structure on the boiling operations of the titanium alloy Ti-6AI-4V, which allow for more insight into the formation modes of burrs between layers. Uniform burr types were not shaped without the backplate and fluid at these cutting conditions. The burrless regions resulted directly from the assistance offered by the backplate to the workpiece. The exit burr sizes were found to be significantly smaller than in the Kim experiment without cutting fluid or a board.

In order to minimize the burr thickness, Kim et al. [10] and Min et al. [18] have developed analytical plots for choosing acceptable clipping conditions in stainless steel AISI 304L and in steel alloy steel materials AISI 4118. They built control charts for prediction by split point twist boring of form and scale of burr. The definition of resemblance was found to be one of the two criteria used for the map. The other was the phase cutting velocity predictor. The diagram was seen to determine form and size of burr, even though the drill diameter varies with feed rate and spindle speed. The geometry of the drill and the material have been set, and the process requirements and the drill diameter were essential parameters. Two control parameters were used for boiling spindle rpm. The usage of these control charts is restricted to drilling across small drilling ranges using a single-layered substrate. Kim and Dornfeld [69] were used to define the process parameters which would govern burr size by means of control charts and Bayes principle. Riech-Weiser et al. [70] formed the third dimension of the burr control map based on earlier observations of material properties and burr shape. Ideally, a dimensionless number should be produced which depends on the material properties influencing the formation of burrs.

Machinery for Lin and Shyu [71] adapted variable feed to increase the existence and exit burr height of the machine for rough and fast to work products. Four coated boxes have been checked, and tests have shown that TIN- and TiCN-coated boxes are more adaptable to boiling steel than the CrN- and TiALN-coated boxes. The creation of burrs in the crushing, heating, milling, graving, or turning industries is one of the main problems of accuracy, industrial processing, and development. The consistency and accuracy of goods are compromised by several forms of burrs which incur additional deburring costs. Deburring, though, accounts for a large portion of manufacturing costs. Therefore, reducing is quite necessary. In this chapter, we research the development of burrs in the inlet and outlet of the hole in the production stage.

8 Important Significance of Burr Formation

Precision components require greater consideration for both the creation of surfaces and measurements such as the diameter with very close tolerances in modern production processes. High quality and precise goods can be manufactured with low manufacturing and processing costs according to the design aspect. To accomplish these aims, it is important to recognize the production process and reduce its parameters. Like machining, metal cutting, injection molding and painting, traditional manufacturing techniques. Drilling for several various machining techniques is one of the most difficult and complex operations. It is commonly employed in many fields, including the aerospace and automotive industries. The projection of material described as "Burr" is one of the biggest dimensional and size errors during drilling. Part of the cycle of machining creates burrs. Burr is a substance extended after processing from the edges or surfaces of the object. It is a deformed substance that is created by cutting or shearing on the edge of component. While boiling, these burrs were noticed when the chisel edge falls in contact with the workpiece at the beginning of the cutting. Once the drill is inserted into the workpiece, the burrs are formed and propagated circumferentially. Undeniably, burr work is rather nuanced. A significant issue is the lack of knowledge of burrs and the process for the creation of burrs in aviation and automobile development cultures. Machining processes also contribute to edge imperfections on the manufactured component in the manufacturing setting. These flaws in the edge may be identified as burrs in the shape of an excellent, rubble substance. In keeping with their teaching method, Gillespie and Blotter have divided the job burrs into four types: birds, rolling over, tear, and cut-off burrs. The rating is based on their training process. The burr in seafood or tension burr stems from the inclination of the substance to bulge when squeezed on the edges. The substance is distorted in the event of a fish burr until there is persistent plastic deformation. The most popular form of burr is a burr, especially a chip, that is bent rather than sliced. Around the end of a break is the broad burr. Breach is the product of material breaking loose rather than cutting from the workpiece. The cut-off burr is triggered by the removal of the component from the raw material prior to completing the removal. Fish mode is observed to predominate for the entrance burrs, while roll over and tear mode are prevalent for the drilling exit burrs.

The three key processes involved in the development of burrs, Gillespie and Blotter, have also been identified: lateral material deformation, chip bending, and chip breaking. An additional, yet related classification method to optimize process parameters such as coated deposition, spindle size, and feed rate parameters with multiple characteristics such as machine existence, surface ruggedness, and burr height was suggested by Nakajama and Aral. It was shown that this approach increased various output characteristics. Boils in different types, such as general carbide boilers, circular drills, chamfer boilers and phase boilers were conducted by Ko et al. [73] drilling research. Burrs were produced with various components, such as steel and aluminum alloys, in specific cutting conditions. Chamber box, circular box, and phase box produces smaller burrs than the traditional unmodified box. As a consequence of the tests, step-boxes of a certain phase angle and phase scale for burr minimization are proposed. Wada and Yoshida [76] highlighted the burrless perforation of specific metals. The curvature of the corner of the drill raising the burr to a very limited amount. The method embraced by Adachi et al. [77] includes modifying the drilling procedure, utilizing ultrasound methods to minimize burr forming. The burr size created by the low-frequency vibrational aluminum boiling was found to be smaller than traditional boiling. However, an analysis was carried out on the connection between the burr size and cutting force and the effects of the cutting force on burr size. Compared with that of carbon coal, the thickness of the burr shaped on aluminum. Figure 11 provides the flowchart for applying the GA in order to optimize the process parameters of drilling. Mainly, eight process input parameters are provided while doing the optimization of the output variables—in the case of drilling burrs, the input parameters are as follows:

- Cutting speed
- Feed
- Use of coolant
- Point geometry
- Point angle
- Lip clearance angle
- Helix angle
- Temperature dependence properties

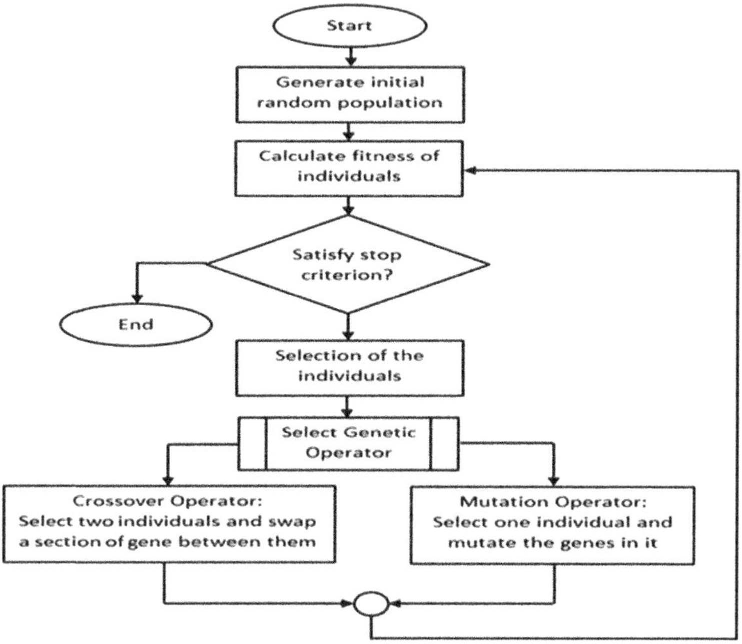

Fig. 11 Flowchart for the process of applying GA to the input parameters

The ultimate aim is to minimize the surface roughness and maximize the precision/accuracy during drilling; in course of this, the readers can use these process parameters and could optimize these input parameters in order to get the desired output.

9 Conclusion and Remarks

Several studies suggest that drill structure, structural properties, and conditions in drilling burr formation are among the main parameters. The cuts in pace and feed were recorded by the majority of studies on burr forming system as influencing process parameters. On the contrary, few research studies have reported that drill diameters, point angles, or lip clearance angles have a major impact on burr shape. Neither the primary nor the interaction results were explored by considering all of the five process parameters, namely pace drilling, feed, drill width, dot angle, and lip angle on burr and burr density, concurrently. Older experiments of burr reduction in potting are known as the target of optimization of either burr height or burr thickness. There were, however, no claims that burr height and burr thickness were decreased simultaneously during boiling.

Genetically dependent process optimization and reliable Taguchi designs are important areas that fulfill the needs of problem solving and product optimization economically. The following assumptions would be taken from the aforementioned findings to articulate the goals of the project. Optimum cutting pace, feed feeder, point angle, and lip clearance angle settings must be defined for a specified drill diameter to reduce burr height and burr thickness simultaneously. A multifaceted optimization method needs simultaneous minimization of burr height and burr thickness. Therefore, the present studies have found approaches focused on genetic algorithms and powerful Taguchi designs. The genetically engineered multifocal optimization approach includes detailed burr height and burr thickness models. Modeling of procedures utilizing the central rotatable composite configuration of the experiments by the response surface methodology (RSM) has proved to be an effective modeling technique. This not only decreases expense and energy; it also includes the requisite details on the key consequences and connections.

The complex modifications proposed to the flexible Taguchi specification are incredibly complicated for multi-response optimization issues. Much of the changes introduced use weighting criteria that are chosen depending on the techniques of testing and mistake. Therefore, a basic improvement to the Taguchi methodology must be made in order to maximize several responses. It could take some time until all burr forming can be stopped during the mechanical part machining. In the meantime, however, a great deal can be accomplished with the technologies and systems outlined in this chapter in order to manufacture components with better efficacy.

References

1. Stein, J. M. (1997). *The buns from drilling: An introduction to drilling burr technology*. Burr Technology Information Series.
2. Gillespie, L. K. (1975). The $2 billion deburring bill. *Manufacturing Engineering and Management, 74*(2), 20–21.
3. Gillespie, L. K. (1999). *Deburring and edge finishing handbook*. Society of Manufacturing Engineers.
4. Ko, S. L., & Dornfeld, D. A. (1991). A study on burr formation mechanism. *Journal of Engineering Materials and Technology, 113*, 75–87.
5. Ko, S. L., & Dornfeld, D. A. (1996). Analysis of fracture in burr formation at the exit stage of metal cut. *Journal of Materials Processing Technology, 58*, 189–200.
6. Ko, T., Yamamoto, A., Kitajima, K., Tanaka, Y., & Takazawa, K. (1991). 'Study on mechanism of burr formation in drilling: Deformation of material during burr formation. *Journal of the Japan Society for Precision Engineering, 57*(3), 485–490.
7. Nakayama, K., & Arai, M. (1987). Burr formation in metal cutting. *Annals CIRP, 36*(1), 33–36.
8. Gillespie, L. K., & Blotter, F. T. (1976). The formation and properties of machining burrs. *Transactions of ASME, Journal of Engineering for Industry, 98*(2), 66–74.
9. Hashimura, M., Ueda, K., & Dornfeld, D. A. (1995). Analysis of three-dimensional burr formation in oblique cutting. *Annals CIRP, 44*(1), 27–30.
10. Kim, J., Min, S., & Dornfeld, D. A. (2001). Optimization and control of drilling burr formation of MSI 3041 and AISI 4118 based on drilling burr control charts. *International Journal of Machine Tools and Manufacture, 41*, 923–936.

11. Kim, J., & Dornfeld, D. A. (2001). Cost minimization of drilling operation by a drilling burr control chart and Bayesian statistics. *SME Journal of Manufacturing Systems, 20*(2), 89–97.
12. Gillespie, L. K. (1975). *Burrs produced by drilling*. Bendix Corporation.
13. Dornfeld, D. A., Kim, J., Dechow, H., Hewson, J., & Chen, L. J. (1999). Drilling burr formation in titanium alloy, Ti-6AI-4V. *Annals CIRP, 48*(1), 73–76.
14. Ko, L., & Lee, J. (2001). Analysis on burr formation in drilling with new concept drill. *Journal of Materials Processing Technology, 113*, 392–398.
15. Kitajima, K., Yamamoto, A., Miyake, T., & Takazawa, K. (2005). Influence of workpiece temperature on burr formation in drilling. *Journal of the Japan Society for Precision Engineering, 71*(2), 252–256.
16. Koelsch, J. (2001). Divining edge quality by reading the burrs. *Quality Magazine, 2001*, 24–28.
17. Arai, M., & Nakayama, K. (1986). Boundary notch on cutting tool caused by burr and its suppression. *Journal of the Japan Society for Precision Engineering, 52*(4), 864–866.
18. Min, S., Kim, J., & Dornfeld, D. A. (2001). Development of a drilling burr control chart for low alloy steel, AISI 4118. *Journal of Materials Processing Technology, 113*, 4–9.
19. Avila, M. C., Choi, J., Dornfeld, D. A., Kapgan, I. M., & Kosarchuk, R. (2004). Deburring of cross-drilled hole intersections by mechanized cutting. In *LMA annual research reports, 2003-2004* (pp. 10–20). LMA.
20. Bakkal, M., Shih, A. J., Samuel, B., McSpadden, Liu, C. T., Ronald, O., & Scattergood. (2005). Light emission, chip morphology and burr formation in drilling the bulk metallic glass. *International Journal of Machine Tools and Manufacture, 45*, 741–752.
21. Schafer, F. (1975). Deburring processes in perspective. In *SME technical papers, MR 75-482*. SME.
22. Choi, H. Z., Lee, S. W., Choi, Y. J., Kim, G. H., & Ko, S. (2004). Micro deburring technology using ultrasonic vibration with abrasive. In *LMA annual research reports, 2003-2004* (pp. 37–43). LMA.
23. Choi, I., & Kim, J. (1998). Electro chemical deburring system using electroplated CBN wheels. *International Journal of Machine Tools and Manufacture, 38*, 29–40.
24. Choi, I., & Kim, J. (1998). A study of the characteristics of the electro chemical deburring of a governor-shaft cross-hole. *Journal of Materials Processing Technology, 75*, 198–203.
25. Dornfeld, D. A., & Erickson, E. (1989). Robotic deburring with real time acoustic emission feedback control. In *Proceedings of symposium on mechanics of debarring and surface finishing processes, ASME winter annual meeting* (pp. 13–26). ASME.
26. Dornfeld, D. A. (1992). Intelligent deburring of precision components. In *Proceedings of IEEE international conference on industrial electronics, control, instrumentation and automation* (pp. 953–960). IEEE.
27. Dornfeld, D. A., & Lisiewicz, V. (1992). Acoustic emission feedback for precision deburring. *Annals CIRP, 41*(1), 93–96.
28. Gillespie, L. K. (1975). *Hand deburring of precision parts*. Bendix Corporation.
29. Gillespie, L. K. (1978). *Advances in deburring*. Society of Manufacturing Engineers.
30. Ko, S. L. (2004). Development of effective measurement system for burr geometry. In *Proceedings of international conference on deburring and surface finishing*. University of California.
31. Lee, S. H., & Dornfeld, D. A. (2001). Precision laser deburring and acoustic emission feedback. *Transactions of ASME, Journal of Manufacturing Science and Engineering, 123*(1), 356–364.
32. Lee, S. H., & Dornfeld, D. A. (2001). Precision laser deburring. *Transactions of ASME, Journal of Manufacturing Science and Engineering, 123*(4), 601–608.
33. Narayanaswami, R., & Dornfeld, D. A. (1994). Design and process planning strategies for burr minimization and deburring. *Transactions of NAMRI/SME, 22*, 313–322.
34. Takazawa, K., Miyatani, T., & Harada, M. (1981). Newly developed deburring machines for precision parts in mass production. In *SME technical papers, MR 81-382*. SME.

35. Yeo, S. H., Bryan, B. K., Ngoy, A., & Chua, L. Y. (1997). Ultrasonic deburring. *The International Journal of Advanced Manufacturing Technology, 13*, 333–341.
36. DeLitzia. (1986). Mechanical deburring with centrifugal blast equipment. *Advancement in Surface Treatment Technology, 2*, 241–254.
37. Kittredge, J. B. (1989). Vibratory finishing equipment. In *SME technical papers, MR89-149*. SME.
38. April, A., & Alwerfalli, D. R. (1975). Deburring metal parts. In *American machinists* (pp. 55–62). American Welding Society.
39. Sonego, R. A. (1988). Electrolyte deburring. *Products Finishing, 53*(2), 57–62.
40. Lee, K., & Dornfeld, D. A. (2002). An experimental study on burr formation in micro milling aluminum and copper. *Transactions of NAMRI/SME, 30*, 255–262.
41. Lee, K., & Dornfeld, D. A. (2005). Micro-burr formation and minimization through process control. *Precision Engineering, 29*(2), 246–252.
42. Leis, K. (2001). Burr formation in drilling intersecting holes. In *LMA annual research reports, 2000-2001* (pp. 13–14). LMA.
43. Kim, J., Dornfeld, D. A., & Furness, R. (1999). Experimental study of burr formation in drilling of intersection holes with gun and twist drills. *Transactions of NAMRI/SME, 27*, 39–44.
44. Min, S., Dornfeld, D. A., & Nakao, Y. (2003). Influence of exit surface angle on drilling burr formation. *Transactions of ASME, Journal of Manufacturing Science and Engineering, 125*(4), 637–644.
45. Stein, J. I. V. I., Park, I. W., & Dornfeld, D. A. (1996). Influence of workpiece exit angle on burr formation in drilling intersecting holes. *Transactions of NAMRI/SME, 24*, 39–44.
46. Gillespie, L. K. (1979). Deburring precision miniature parts. *Precision Engineering, 1*(4), 189–198.
47. Gillespie, L. K. (1981). *Deburring technology for improved manufacturing*. Society of Manufacturing Engineers.
48. Gillespie, L. K. (1996). *Standard terminology for researchers of burrs and edge finishing*. Worldwide Burr Technology Committee.
49. Sickle, C. V., & Flores, G. V. (1997). How to pick the right deburring process. *Manufacturing Engineering, 1997*, 56–62.
50. Gillespie, L. K. (1976). *Effects of drilling variables on burr properties*. Bendix Corporation.
51. Sofronas, A., Spurgeon, M., & Taraman, K. (1975). Reduction of burr formation in drilling. In *SME technical papers, MR 75-376*. SME.
52. Stein, J. M., & Dornfeld, D. A. (1997). Burr formation in drilling miniature holes. *Annals CIRP, 46*(1), 63–67.
53. Chern, G. L., & Dornfeld, D. A. (1996). Burr/breakout model development and experimental verification. *Transactions of ASME, Journal of Engineering Materials and Technology, 118*(2), 201–206.
54. Sofronas, A., & Taraman, K. (1976). Model development for exit burr thickness as a function of drill geometry and feed. In *SME technical papers, MR 76-253*. SME.
55. Shikata, H., DeVries, M. F., & Wu, S. M. (1980). An experimental investigation of sheet metal drilling. *Annals CIRP, 29*(1), 85–88.
56. Sofronas, A. (1976). The effect of system stiffness, workpiece hardness and spindle speed on drilling burr thickness. In *SME technical papers, MR 75-132*. SME.
57. Sugawara, A., & Inagaki, K. (1978). Effect of shape of tool point with dwindling of drill diameter on drilling: Burr in case of 0.02% of C steel. *Journal of the Japan Society for Precision Engineering, 44*(2), 179–184.
58. Sugawara, A., & Inagaki, K. (1981). Burr in micro diameter drill working. *Journal of the Japan Society for Precision Engineering, 15*(1), 21–26.
59. Min, S., Lee, D., & Grave, A. (2005). Surface and edge quality variation in precision machining of single crystal and polycrystalline materials. In *LMA annual research reports, 2003-2004* (pp. 52–62). LMA.

60. Takazawa, K. (1988). The challenge of burr technology and its worldwide trends. *Journal of the Japan Society for Precision Engineering, 22*(3), 165–170.
61. Kim, J. (1998). Preliminary experiment of drilling burr formation in titanium alloy. In *LMA annual research reports, 1998-1999* (pp. 33–34). LMA.
62. Kim, J., & Dornfeld, D. A. (2002). Development of an analytical model for drilling burr formation in ductile materials. *Transactions of ASME, Journal of Engineering Materials and Technology, 124*, 192–198.
63. Kim, J. (1999). Probabilistic approach of burr type prediction with the drilling burr control chart. In *LMA annual research reports, 1999-2000* (pp. 42–43). LMA.
64. Heisel, U., Luik, M., Eisseler, R., & Schaal, M. (2005). Prediction of parameters for the burr dimensions in short - hole drilling. *Annals CIRP, 54*(1), 79–83.
65. Shefelbine, W., & Dornfeld, D. A. (2004). The effect of dry machining on burr size. In *LMA annual research reports, 2003-2004* (pp. 71–75). LMA.
66. Pande, S. S., & Relekar, H. P. (1986). Investigations on reducing burr formation in drilling. *International Journal of Machine Tool Design and Research, 26*(3), 339–348.
67. Sugawara, A., & Inagaki, K. (1982). Effect of workpiece structure on burr formation in micro drilling. *Precision Engineering, 4*(1), 9–14.
68. Hewson, J. (1999). Exit burr size and shape in backplate assisted drilling of Ti-6AI-4V. In *LMA annual research reports, 1998-1999* (pp. 35–36).
69. Kim, J., & Dornfeld, D. A. (2000). Development of a drilling burr control chart for stainless steel. *Transactions of NAMRI/SME, 28*, 317–322.
70. Riech-Weiser, C., & Dornfeld, D. A. (2005). Drilling burr control chart - adding a material property axis. In *LMA annual reports, 2004-2005* (pp. 19–21). LMA.
71. Lin, T. R., & Shyu, R. F. (2000). Improvement of tool life and exit burr using variable feeds when drilling stainless steel with coated drills. *The International Journal of Advanced Manufacturing Technology, 16*, 308–313.
72. Huang, M. F., & Lin, T. R. (2004). Application of grey –taguchi method to optimise drilling of aluminium alloy 6061 with multiple performance characteristics. *Materials Science and Technology, 20*(4), 528–532.
73. Ko, S. L., Chang, J. E., & Kalpakjian, S. (2003). Development of drill geometry for burr minimization in drilling. *Annals CIRP, 52*(1), 45–48.
74. Jean, M. D., & Wang, J. T. (2006). Using a principal components analysis for developing a robust design of electron beam welding. *The International Journal of Advanced Manufacturing Technology, 28*, 882–889.
75. Ko, S. L., Chang, J. E., & Yang, G. E. (2003). Burr minimizing scheme in drilling. *Journal of Materials Processing Technology, 140*, 237–242.
76. Wada, H., & Yoshida, K. (2000). Burrless drilling of metals. *Journal of the Japan Society for Precision Engineering, 66*(7), 1109–1114.
77. Adachi, K., Arai, N., Harada, S., Okita, K., & Wakisaka, S. (1987). A study on burr in low frequency vibratory drilling. *Journal of the Japan Society for Precision Engineering, 21*(4), 258–264.
78. Takeyama, H., & Kato, S. (1991). Burrless drilling by means of ultrasonic vibration. *Annals CIRP, 40*(1), 83–86.
79. Takeyama, H. (1993). Study on oscillatory drilling aiming at prevention of burr. *Journal of the Japan Society for Precision Engineering, 59*(10), 1719–1724.
80. Simon, F. C., & Bone, G. M. (2005). Burr size reduction in drilling by ultrasonic assistance. *Robotics and Computer-Integrated Manufacturing, 21*, 442–450.
81. Lee, K. (2002). Optimization and quality control in burr formation using design of experiment (II): Drilling intersecting holes. In *LMA annual research reports, 2001-2002* (pp. 45–47). LMA.
82. Tosun, N. (2006). Determination of optimum parameters for multi-performance characteristics in drilling by using grey relational analysis. *The International Journal of Advanced Manufacturing Technology, 28*, 450–455.

83. Iwata, K., Ueda, K., & Okuda, K. (1982). Study of mechanism of burrs formation in cutting based on direct SEM observations. *Journal of the Japan Society for Precision Engineering, 48*(4), 510–515.
84. Guo, Y. B., & Dornfeld, D. A. (1998). Integration of CAD of drill with FEA of drilling burr formation. *Transactions of NAMRI/SME, 26*, 201–206.
85. Guo, Y. B., & Dornfeld, D. A. (1998). Finite element analysis of drilling burr minimization with a backup material. *Transactions of NAMRI/SME, 26*, 207–212.
86. Park, I. W., & Dornfeld, D. A. (2000). A study on burr formation processes using the finite element method - Part I. *Transactions of ASME, Journal of Engineering Materials and Technology, 122*, 221–228.
87. Park, I. W., & Dornfeld, D. A. (2000). A study on burr formation processes using the finite element method - Part II - The influences of exit angle. Rake angle and backup material on burr formation processes. *Transactions of ASME, Journal of Engineering Materials and Technology, 122*, 229–237.
88. Min, S. (1998). Finite element modeling of burr formation in 2-D orthogonal cutting with a backup material. In *LMA annual research reports, 1998-1999* (pp. 39–40). LMA.
89. Min, S., Dornfeld, D. A., Kim, J., & Shyu, B. (2001). Finite element modeling of burr formation in metal cutting. *Machining Science and Technology, 5*(2), 307–322.
90. Guo, Y. B., & Dornfeld, D. A. (2000). Finite element modeling of burr formation process in drilling 304 stainless steel. *Transactions of ASME, Journal of Manufacturing Science and Engineering, 122*(4), 612–619.
91. Choi, J., Iin, S., Dornfeld, D. A., Mehboob, A., & Tzong, T. (2003). Modeling of interlayer gap formation in drilling of a multi-layered material. In *LMA annual research reports, 2002-2003* (pp. 36–41). LMA.
92. Choi, J., Min, S., & Dornfeld, D. A. (2004). Finite element modeling of burr formation in drilling of a multi-layered material. In *LMA annual research reports, 2003-2004* (pp. 31–36). LMA.
93. Vijayaraghavan, A., & Gardner, J. D. (2005). Comparative study of finite element simulation software. In *LMA annual research reports, 2004-2005* (pp. 15–18). LMA.
94. Vijayaraghavan, A. (2005). Challenges in modeling machining of multi-layer materials. In *LMA annual research reports, 2004-2005* (pp. 30–36). LMA.
95. Peria, B., Gorka, A., Rivero, & Lopez de Lacalle, L. (2005). Monitoring of drilling for burr detection using spindle torque. *International Journal of Machine Tools and Manufacture, 45*, 1614–1621.
96. Box, G. E. P., & Draper, N. R. (1987). *Empirical model-building and response surfaces*. John Wiley & Sons.
97. Khuri, A. I., & Cornell, J. A. (1996). *Response surfaces: Design and analyses*. Marcel Dekker.
98. Myers, R. H., & Montgomery, D. C. (1995). *Response surface methodology: Process and product optimization using designed experiments*. Wiley.
99. Simpson, T. W., Dennis, L., & Chen, C. (2002). Sampling strategies for computer experiments: Design and analysis. *International Journal of Reliability and Applications, 23*(2), 209–240.
100. Feng, C. X., & Wang, X. (2002). Development of empirical models for surface roughness prediction in finish turning. *The International Journal of Advanced Manufacturing Technology, 20*, 348–356.
101. Huang, L., & Chen, J. C. (2001). A multiple regression model to predict in-process surface roughness in turning operation via accelerometer. *Journal of Industrial Technology, 17*(2), 1–8.
102. Manna, A., & Bhattacharya, B. (2004). Investigation for optimal parametric combination for achieving better surface finish during turning of Al/SiC-MMC. *The International Journal of Advanced Manufacturing Technology, 23*, 658–665.

103. Noordin, M. Y., Venkatesh, V. C., Sharif, S., Elting, S., & Abdullah, A. (2004). Application of response surface methodology in describing the performance of coated carbide tools when turning AISI 1045 steel. *Journal of Materials Processing Technology, 145*, 46–58.
104. Singh, D., & Rao, P. V. (2007). A surface roughness prediction model for hard turning process. *International Journal of Advanced Manufacturing Technology, 32*(11), 1115–1124.
105. Onwubolu, G. C., & Kumar, S. (2006). Response surface methodology - Based approach to CNC drilling operations. *Journal of Materials Processing Technology, 171*, 41–47.
106. Davim, J. P., & Reis, P. (2004). Multiple regression analysis (MRA) in modeling milling of glass fibre reinforced plastics (GFRP). *International Journal of Manufacturing Technology and Management, 6*(2), 185–197.
107. Lou, M. S., Chen, J. C., & Li, M. C. (1998). Surface roughness prediction technique for CNC end-milling. *Journal of Industrial Technology, 15*(1), 1–6.
108. Bhattacharya, B., & Sorkhel, S. K. (1999). Investigation for electro chemical machining through response surface methodology based approach. *Journal of Materials Processing Technology, 86*, 200–207.
109. Karthikeyan, R., Lakshmi Narayan, P. R., & Nagarazan, R. S. (1999). Mathematical modeling for electric discharge machining of aluminum - silicon carbide particulate composites. *Journal of Materials Processing Technology, 87*, 59–63.
110. Sarkar, B. R., Doloi, B., & Bhattacharyya, B. (2006). Parametric analysis on electrochemical discharge machining of silicon nitride ceramics. *The International Journal of Advanced Manufacturing Technology, 28*, 873–881.
111. Sen, M., Shan, H., & S. (2006). Response surface analysis of electro jet drilled holes. *International Journal of Advanced Manufacturing Technology, 31*, 520–527.
112. Singh, D. K., Jain, V. K., & and- Raghuram, V. (2006). Experimental investigations into forces acting during a magnetic abrasive finishing process. *The International Journal of Advanced Manufacturing Technology, 30*, 652–662.
113. Gunaraj, V., & Murugan, N. (1999). Application of response surface methodology for predicting weld bead quality in submerged arc welding of pipes. *Journal of Materials Processing Technology, 88*, 266–275.
114. Gunaraj, V., & Murugan, N. (2002). Prediction of heat-affected zone characteristics in submerged arc welding of structural steel pipes. *Welding Journal, 2002*, 94–98.
115. Murugan, N., Parmar, R. S., & Sud, S. K. (1993). Effect of submerged arc process variables on dilution and bead geometry in single wire surfacing. *Journal of Materials Processing Technology, 37*, 767–780.
116. Chen, D. C., Hsu, R., & Fuh, K. (2005). Effect of over-roll thickness on cone surface roughness in shear spinning. *Journal of Materials Processing Technology, 159*, 1–8.
117. Tiernan, P., Draganescu, B., & Hillery, M. T. (2005). Modelling of extrusion force using the surface response method. *The International Journal of Advanced Manufacturing Technology, 27*, 48–52.
118. Papila, M., & Haftka, R. T. (2000). Response surface approximations: Noise, error repair and modeling errors. *AIAA Journal, 38*(12), 2336–2343.
119. Hussain, M. F., Barton, R. R., & Joshi, S. B. (2002). Metamodeling: Radial basis functions versus polynomials. *European Journal of Operational Research, 138*(1), 142–154.
120. Klir, G. J., & Yuan, B. (1998). *Fuzzy system and fuzzy logic -Theory and practice*. Prentice Hall.
121. Tong, L. I., & Su, C. T. (1997). Optimizing multi-response problems in the taguchi method by fuzzy multiple attribute decision making. *Quality and Reliability Engineering International, 13*, 25–34.
122. Haykin, S. (1999). *Neural networks: A comprehensive foundation*. Prentice-Hall.
123. Simpson, T. W., Peplinski, J. D., Koch, P. N., & Allen, J. K. (2001). Metamodels for computer-based engineering design: Survey and recommendations. *Engineering with Computers, 17*(2), 129–150.
124. Powell, M., & J. D. (1987). *Radial basis functions for multivariable interpolation: A review*. Oxford University Press.

125. Jin, R., Chen, W., & Simpson, T. W. (2001). Comparative studies of metamodelling techniques under multiple modelling criteria. *Structural and Multidisciplinary Optimization, 23*(1), 1–13.
126. Sakata, A., Ashida, F., & Zako, M. (2003). Structural optimization using kriging approximation. *Computer Methods in Applied Mechanics and Engineering, 192*(7-8), 923–939.
127. Buhmann, M. D. (2000). Radial basis functions. *Acta Numerica, 9*, 1–38.
128. Chang, S. I., & Shivpuri, R. (1995). A multiple-objective decision making approach for assessing simultaneous improvement in die life and casting quality in a die casting process. *Journal of Quality Technology, 29*, 339–346.
129. Joseph, J., & Pignatiello, J. R. (1993). Strategies for robust multi response quality engineering. *Transactions on Industrial Engineering Research Development, 25*(3), 5–15.
130. Cohon, J. L. (1985). *Multi criteria programming: Brief review and application in design optimization*. Academic Press.
131. Charnes, A., & Cooper, W. W. (1961). *Management models and industrial applications of linear programming* (Vol. 7). Wiley.
132. Goldberg, D. E. (1989). *Genetic algorithms in search optimization and machine learning*. Addison Wesley Publishing Company Inc..
133. Antony, J. (2001). Simultaneous optimisation of multiple quality characteristics in manufacturing processes using Taguchi's quality loss function. *The International Journal of Advanced Manufacturing Technology, 17*, 134–138.
134. Reddy, N. S., & Rao, P. V. (2005). Selection of optimum tool geometry and cutting conditions using a surface roughness prediction model for end milling. *The International Journal of Advanced Manufacturing Technology, 26*, 1202–1210.
135. Reddy, N. S., & Rao, P. V. (2006). Selection of an optimal parametric combination for achieving a better surface finish in dry milling using genetic algorithms. *The International Journal of Advanced Manufacturing Technology, 28*, 463–473.
136. Saravan, R. (2002). A multi objective genetic algorithm (GA) approach for optimization of surface grinding. *International Journal of Machine Tools and Manufacture, 42*, 1327–1334.
137. Suresh, P. V. S., Rao, P. V., & Deshmukh, S. G. (2002). A genetic algorithm approach for optimization of surface roughness prediction model. *International Journal of Machine Tools and Manufacture, 42*, 675–680.
138. Wang, X., & Jawahir, I. S. (2004). Web based optimization of milling operations for the selection of cutting conditions using genetic algorithms. *Proceedings of Institute of Mechanical Engineers, 218*, 212–223.
139. Wen, J., Cheng, S., & Malik, O. P. (1998). A synchronous generator fuzzy excitation controller optimally designed with a genetic algorithm. *IEEE Transactions on Power Systems, 13*(3), 884–889.
140. Ghanl, A. J., Choudhury, A. I., & Hassan, H. H. (2004). Application of Taguchi method in the optimization of end milling parameters. *Journal of Materials Processing Technology, 145*, 84–92.
141. Khoo, L. P., & Chen, C. H. (2001). Integration of response surface methodology with genetic algorithms. *The International Journal of Advanced Manufacturing Technology, 18*, 483–489.
142. Deb, K. (1995). *Optimization for engineering design: Algorithms and examples*. Prentice-Hall.
143. Dorigo, I. V. I. (1996). The ant system: Optimization by a colony of cooperating agent. *IEEE Transaction Systems on Man Cybernetics, 26*(1), 1–13.
144. Glover, F. (1999). *Scatter search and path relinking - New ideas in optimization*. McGraw-Hill.
145. Glover, F., Laguna, M., & Marti, R. (2000). Fundamentals of scatter search and path relinking. *Control and Cybernetics, 39*(3), 653–684.
146. Chen, M. C., & Chen, K. Y. (2003). Determination of optimum machining conditions using scatter search. *New Optimization Techniques in Engineering, 2*, 681–697.
147. Chen, M. C. (2004). Optimizing machining economics models of turning operations using the scatter search. *International Journal of Production Research, 42*(13), 2611–2625.

148. Krishna, A. G., & Rao, K. M. (2006). Multi optimization of surface grinding operations using scatter search approach. *The International Journal of Advanced Manufacturing Technology, 29*, 475–480.
149. Kennedy, J., & Eberhart, R. (1995). Particle swarm optimization. In *Proceedings of IEEE international conference on neural networks, Part IV* (pp. 942–948). IEEE.
150. Christu, P. R., Asokan, P., & Prabhakar, V. I. (2006). A solution to the facility layout problem having passages and inner structure walls using particle swarm optimization. *The International Journal of Advanced Manufacturing Technology, 29*, 766–771.
151. Karpat, Y., & Ozel, T. (2005). Hard turning optimization using neural network modeling and swarm intelligence. *Transactions of NAMRI/SME, 33*, 179–186.
152. Natarajan, U., Saravanan, R., & Periasamy, V. M. (2006). Application of particle swarm optimisation in artificial neural network for the prediction of tool life. *The International Journal of Advanced Manufacturing Technology, 28*, 1084–1088.
153. Bagchi, T. P. (1993). *Taguchi methods explained - Practical steps to robust design*. Prentice-Hall of India.
154. Logothetis, N., & Haigh, A. (1988). Characterizing and optimizing multi response processes by the Taguchi method. *Quality and Reliability Engineering International, 4*(2), 159–169.
155. Phadke, M. (1989). *Quality engineering using robust design*. Prentice Hall.
156. Taguchi, G. (1990). *Introduction to quality engineering*. Asian Productivity Organization.
157. Davim, J. P. (2003). Design of optimization cutting parameters for turning metal matrix composites based on tile orthogonal arrays. *Journal of Materials Processing Technology, 132*, 340–344.
158. Deng, C. S., & Chin, J. H. (2005). Hole roundness in deep hole drilling as analysed by Taguchi methods. *The International Journal of Advanced Manufacturing Technology, 25*, 420–426.
159. GhanI, A. J., Choudhury, A. I., & Hassan, H. H. (2004). Application of Taguchi method in the optimization of end milling parameters. *Journal of Materials Processing Technology, 145*, 84–92.
160. Kopac, J., Behor, M., & Sokovic, M. (2002). Optimal machining parameters for achieving the desired surface roughness in fine turning of cold pre-formed steel workpieces. *International Journal of Machine Tools and Manufacture, 42*, 707–716.
161. Liu, S., & Chen, C. (2000). Optimization of bubble size in rotationally moulded parts. *Plastics, Rubber and Composites, 29*(8), 411–418.
162. Iarafrona, J., & Wyles, C. (2000). A new method of optimizing material removal rate using EDM. *International Journal of Machine Tools and Manufacture, 40*, 153–164.
163. Oktem, H., Erzurumlu, T., & Col, M. (2006). A study of the taguchi optimization method for surface roughness in finish milling of mold surfaces. *The International Journal of Advanced Manufacturing Technology, 28*, 694–700.
164. Shiou, F. J., Chen, C., & H. (2003). Determination of optimal ball-burnishing parameters for plastic injection moulding steel. *The International Journal of Advanced Manufacturing Technology, 3*, 177–183.
165. Yang, J. L., & Chen, J. C. (2001). A systematic approach for identifying optimum surface roughness performance in end milling operations. *Journal of Industrial Technology, 17*(2), 1–8.
166. Jeyapaul, R., Shahabudeen, P., & Krishnaiah, K. (2005). Quality management research by considering multi-response problems in the Taguchi method - A review. *The International Journal of Advanced Manufacturing Technology, 26*, 1331–1337.
167. Tosun, N., & Ozler, L. (2004). Optimisation for hot turning operations with multiple performance characteristics. *The International Journal of Advanced Manufacturing Technology, 23*, 777–782.
168. Yang, W. H., & Tarng, Y. S. (1998). Design optimization of cutting parameters for turning operations based on the Taguchi method. *Journal of Materials Processing Technology, 84*, 122–129.

169. Vining, G. G., & Myers, R. H. (1990). Combining Taguchi and response surface philosophies: A dual response approach. *Journal of Quality Technology, 22*, 38–45.
170. Keeney, R. L., & Raiffa, H. (1976). *Decisions with multiple objectives: Preferences and value tradeoffs*. Wiley.
171. Kumar, P., Barua, P. B., & Gaindhar, J. L. (2000). Quality optimization (multi-characteristic) through Taguchi's technique and utility concept. *Quality and Reliability Engineering International, 16*(6), 475–485.
172. Yang, C., & Hung, S. W. (2004). Optimising the thermoforming process of polymeric foams: An approach by using the Taguchi method and the utility concept. *The International Journal of Advanced Manufacturing Technology, 24*, 353–360.
173. Ames, A. E., Mattucci, N., Macdonald, S., Szonyi, G., & Hawkins, D. M. (1997). Quality loss functions for optimization across multiple response surfaces. *Journal of Quality Technology, 29*, 339–346.
174. Elsayed, E. A., & Chen, A. (1993). Optimal levels of process parameters for products with multiple characteristics. *International Journal of Production Research, 31*(5), 1117–1132.
175. Lin, T. R. (2002). Optimisation technique for face milling stainless steel with multiple performance characteristics. *The International Journal of Advanced Manufacturing Technology, 19*, 330–335.
176. Nian, C. Y., Yang, W. H., & Tarng, Y. S. (1999). Optimization of turning operations with multiple performance characteristics. *Journal of Materials Processing Technology, 95*, 90–96.
177. Ramakrishnan, R., & Karunamoorthy, L. (2005). Multi response optimization of wire EDM operations using robust design of experiments. *The International Journal of Advanced Manufacturing Technology, 29*, 105–112.
178. Tarng, Y. S., & Yang, W. H. (1998). Optimisation of the weld bead geometry in gas tungsten arc welding by the Taguchi method. *The International Journal of Advanced Manufacturing Technology, 14*, 549–554.
179. Tarng, Y. S., & Yang, W. H. (1998). Application of the Taguchi method to the optimisation of the submerged arc welding process. *Materials and Manufacturing Processes, 13*(3), 455–467.
180. Jolliffe, I. T. (1986). *Principal component analysis*. Springer.
181. Su, C. T., & Tong, L. I. (1997). Multi response robust design by principal component analysis. *Total Quality Management, 8*(6), 409–416.
182. Yang, K. (1996). Improving automotive dimensional quality by using principal component analysis. *Quality and Reliability Engineering International, 12*, 401–409.
183. Johnson, D. E. (1998). *Applied multivariate methods for data analysts*. Duxbury.
184. Sharma, S. (1996). *Applied multivariate techniques*. Wiley and Sons.
185. Tong, L. I., Wang, C. H., & Chen, H. C. (2005). Optimization of multiple response using principle component analysis and technique for order preference by similarity to ideal solution. *The International Journal of Advanced Manufacturing Technology, 27*, 407–414.
186. Cabrera, R. M., Castro, J. M., & Mount-Campbell, C. A. (2002). Multiple quality criteria optimization in reactive in mold coating (IMC) with a data envelopment analysis approach. *Journal of Polymer Engineering, 22*(5), 305–340.
187. Castro, C., Cabrera, R. M., Lily, B., & Castro, J. M. (2003). Identifying the best compromises between multiple performance measures in injection molding (IM) using data envelopment analysis (DEA). *Journal of Integrated Design and Process Science, 7*(1), 77–86.
188. Deng, J. L. (1989). Introduction to grey system. *The Journal of Grey System, 1*(1), 1–24.
189. Chang, C.-K., & Lu, H. S. (2007). Design optimization of cutting parameters for side milling operations with multiple performance characteristics. *The International Journal of Advanced Manufacturing Technology, 32*, 18–26.
190. Chang, C. S., Liao, R. C., Wen, K. L., & Wang, W. P. (2004). A grey based Taguchi method to optimize design of muzzle flash restraint device. *The International Journal of Advanced Manufacturing Technology, 24*, 860–864.

191. Huang, J. T., & Liao, Y. S. (2003). Optimization of machining parameters of wire-EDM based on grey relational and statistical analysis. *International Journal of Production Research, 41*(8), 1707–1720.
192. Kao, P. S., & Hocheng, H. (2003). Optimization of electrochemical polishing of stainless steel by grey relational analysis. *Journal of Materials Processing Technology, 140*, 255–259.
193. Kuo, F. J., & Wu, Y. (2006). Optimization of the film coating process for polymer blends by the grey-based Taguchi method. *The International Journal of Advanced Manufacturing Technology, 27*, 525–530.
194. Lin, J. L., & Tarng, Y. S. (1998). Optimization of multi-response process by the Taguchi method with grey relational analysis. *The Journal of Grey System, 10*(4), 355–370.
195. Lin, J. L., & Lin, C. L. (2002). The use of the orthogonal array with grey relational analysis to optimize the electrical discharge machining process with multiple performance characteristics. *Journal of Materials Processing Technology, 42*, 237–244.
196. Lin, Z. C., & Ho, C. Y. (2003). Analysis and application of grey relational and ANOVA in chemical-mechanical polishing process parameters. *The International Journal of Advanced Manufacturing Technology, 21*, 10–14.
197. Tarng, Y. S., Juang, S. C., & Chang, C. H. (2002). The use of grey based Taguchi methods to determine submerged arc welding process parameters in hardfacing. *International Journal of Machine Tools and Manufacture, 128*, 1–6.
198. Derringer, G., & Suicli, R. (1980). Simultaneous optimization of several response variables. *Journal of Quality Technology, 12*, 214–219.
199. Tong, L.-I., Chen, C.-C., & Wang, C.-H. (2007). Optimization of multi-response processes using the VIKOR method. *The International Journal of Advanced Manufacturing Technology, 31*, 1049–1057.

Investigations on Mechanical Properties of Reinforced Secondary Recycled ABS as Filament for 3D Printing Applications

K. Chawla, Rupinder Singh, and J. Singh

1 Introduction

In today's world, two manufacturing techniques have been used to create products: first one is traditional manufacturing (machining) and second is additive manufacturing (AM). AM techniques have the capability to create products having customized features with acceptable accuracy. Also, the traditional machining is time consuming and unable to create complicated 3D parts within stipulated time [1, 2]. One of the remarkable features of AM techniques compared to traditional machining processes is their ability to develop new products with significant reduced time from production to market [3]. There are certain challenges with AM techniques also like restricted to few materials, lack of testing standards for evaluation and requirement of post processing techniques to improve surface integrity [4]. Despite of these limitations, nowadays researcher introduced AM in defence, medical, electronics, art, and entertainment [5]. Among the various available AM techniques, FDM is one of the simplest techniques developed in 1980 [6, 7]. In FDM, different thermoplastic polymers such as ABS, Polyamide, Nylon-6, polycarbonate, poly lactic acid, etc. are utilized as fused filament [8]. ABS is one of the commonly used material in the FDM due to its high wear resistance property, good thermal property and high durability [9]. As ABS is non-biodegradable polymer, this property makes it suitable for non-structural applications. Singh et al. [10] prepared feedstock filament by reinforcing Al_2O_3 in Nylon 6 polymer for investigating the effect of particle

K. Chawla · J. Singh
School of Mechanical Engineering, Lovely Professional University, Phagwara, India

R. Singh (✉)
Department of Mechanical Engineering, National Institute of Technical Teachers Training and Research, Chandigarh, India
e-mail: rupindersingh@nitttrchd.ac.in

C. Prakash et al. (eds.), *Additive, Subtractive, and Hybrid Technologies*,
Mechanical Engineering Series, https://doi.org/10.1007/978-3-030-99569-0_3

size on thermal properties. Billah et al. [11] reported that with the addition of carbon fibre in ABS, thermal properties of composite can be improved up to certain level. Jin et al. [12] investigated that addition of poly(methyl methacrylate) in ABS improves the mechanical properties of the composite. Some of the researchers revealed that strength of thermoplastic composite can be improved up to certain extent with the addition of wood, graphene, nano silica and hollow glass beads in ABS [13–15]. Lohar and Jogi [16] obtained higher tensile and impact strength when carbon black was reinforced in polypropylene and ABS matrix. Some researchers enhanced the thermal conductivity of ABS composite to 0.379 w/m-k by reinforcing carbon fibre [17]. Chawla et al. [18] utilized recycled ABS to prepare filaments for FDM technique without any reinforcement. Throughout the world, research is going on to develop composites for engineering applications with the addition of different materials [19, 20]. It has been observed that most of the researchers used virgin polymer as base material for fabricating 3D printed parts instead of recycled polymer [21, 22]. Chawla et al. [23] successfully fabricated composite filaments based on recycled ABS for non-structural engineering applications.

From literature, it was predicted that very less work has been reported on composite filaments prepared from recycled polymer. Also, most of the researches focused on blending of one or two reinforced materials, but hitherto very less work has been done on hybrid blending of three reinforcements in base material. Therefore, in the present study, composite filaments were fabricated by utilizing 2° recycled ABS as base material reinforced with wood dust, iron (Fe) powder and bakelite powder at weight proportion of 10% (by weight) for FDM process. The in-house hybrid blended composite feed stock filaments were prepared with the help of twin-screw extruder (TSE) and their rheological, thermal and mechanical properties were investigated.

2 Materials Selection and Experimentation

In present study, 2° recycled ABS has been procured from the market, and reinforced materials were collected from various resources such as wood dust taken from saw mill, Fe powder from milling shop and bakelite from metallurgical lab. A standard particle size of 50 μm was selected for all the reinforced materials. After collection of base and reinforced materials, a composite was prepared by taking 90% ABS and 10% reinforcements (bakelite powder + wood dust + Fe powder). Then the MFI (melt flow index) of the prepared composite was determined as per ASTM D1238 standards which indicates the ability of the polymer to flow through nozzle in 10 min under standard conditions. After doing five repeated trials for prepared composite, MFI comes out as 18.35 g/10 min. On further increase in reinforcement proportion clogging of the nozzle took place.

L9 orthogonal array has been applied on TSE for the fabrication of feed stock filament by varying load, temperature and speed as per Table 1. Different mechanical properties such as break strength, peak strength, modulus of elasticity and

Table 1 Control log of experimentation for hybrid filament

Experiment No.	1	2	3	4	5	6	7	8	9
Load (kg)	10	10	10	12.5	12.5	12.5	15	15	15
Temperature (°C)	225	235	245	225	235	245	225	235	245
Speed (rpm)	70	80	90	80	90	70	90	70	80

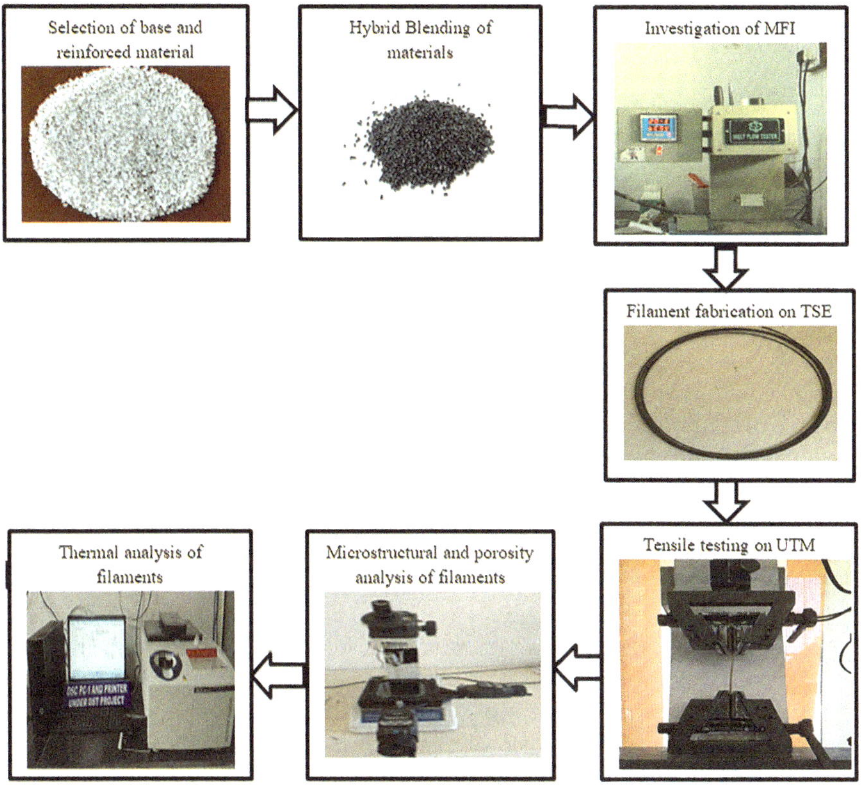

Fig. 1 Methodology used in the present study

toughness of prepared composite filaments have been determined with the help of universal testing machine (UTM). Further, optical photomicrographs, scanning electron microscopy (SEM) images and porosity analysis has been performed on the composite filaments for determining the effect of reinforcement percentage on mechanical properties. In last, differential scanning calorimetry (DSC) has been used for studying the thermal behaviour of filaments. The methodology used in the current study was represented in Fig. 1.

3 Results and Discussion

After preparing the nine different filaments with the help of TSE as per the conditions mentioned in Table 1, tensile testing has been performed on all the filaments with the help of UTM machine. The different mechanical properties obtained for the fabricated filaments were represented in Table 2. A stress–strain curve has been plotted with the obtained values as shown in the Fig. 2. From stress–strain curve, it has been observed that sample 6 possesses maximum toughness 0.550 MPa while sample 3 gives minimum 0.10 MPa among all the filaments.

Optical photomicrographs were taken at the break cross-section for sample no. 8 and 3 having maximum and minimum peak strength (PS) as shown in Fig. 3. From photomicrograph analysis, it has been found that the quantity of voids in all the filaments were not same and filaments fabricated at 225 °C contains relatively higher number of voids as compared to other samples.

Further, both the selected optical photomicrograph images were processed on image processing software for estimating the average surface roughness (R_a) and surface topology. The results obtained for average surface roughness (R_a) in the form of graph has been shown in Fig. 4. From analysis, it has been found that sample 8 has minimum R_a (24.38 nm), whereas sample 3 has maximum R_a (32.43 nm) which could be due to the presence of higher number of voids. Therefore, it has been found that the presence of higher number of asperities/voids in the fabricated filaments could be one of the reasons for showing poor mechanical properties. Tool Maker Microscope has been used to determine the porosity of filaments at ×100 as per standard ASTM B276. The porosity results obtained in the form of image has been shown in Fig. 5. From porosity analysis, it was observed that sample 3 has maximum porosity percentage (17.74%), whereas sample 8 has minimum porosity percentage (4.96%). Due to this reason, sample 8 exhibits poor mechanical properties. Further, signal to noise (S/N) ratio has been calculated for PS based on larger is better approach as represented in Table 3. On the basis of calculated S/N, mean effect plot for PS has been obtained as shown in the Fig. 6. PS is an important parameter affecting the strength of filament; therefore, ANOVA has been applied on

Table 2 Mechanical properties of fabricated composite filaments

S. No	Break strength (MPa)	Peak strength (MPa)	Modulus of elasticity (GPa)	Toughness (MPa)
1	8.6 ± 1.6	9.65 ± 1.8	0.380 + 1.4	0.122 ± 0.15
2	8.54 ± 1.5	9.55 ± 1.6	0.376 ± 0.3	0.107 ± 0.12
3	9.11 ± 0.9	9.21 ± 1.2	0.484 ± 0.4	0.10 ± 0.11
4	10.1 ± 1.2	11.76 ± 1.3	0.53 ± 0.35	0.125 ± 0.14
5	9.51 ± 1.1	10.21 ± 1.2	0.104 ± 0.45	0.466 ± 0.13
6	8.69 ± 1.1	13.79 ± 1.1	1.08 ± 0.3	0.550 ± 0.12
7	11.13 ± 1.0	14.38 ± 1.1	0.648 ± 0.4	0.123 ± 0.13
8	16.29 ± 1.2	20.47 ± 1.4	3.2 ± 0.3	0.21 ± 0.12
9	10.59 ± 1.2	17.89 ± 1.1	0.941 ± 0.35	0.117 ± 0.13

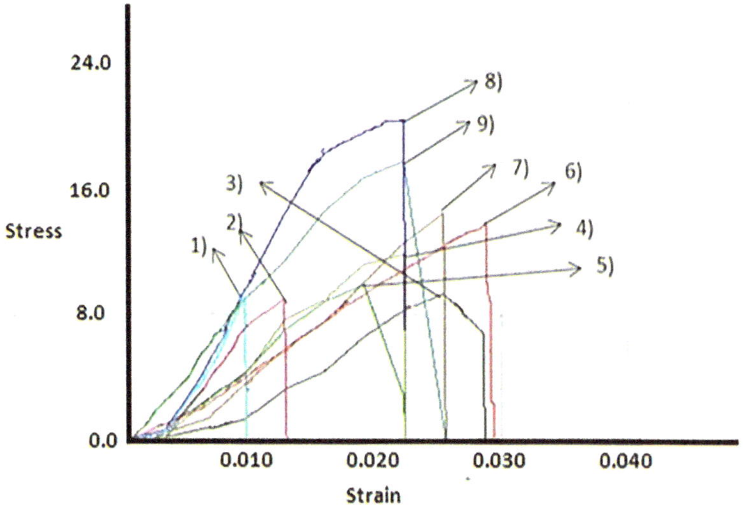

Fig. 2 Stress–strain curve for prepared filaments

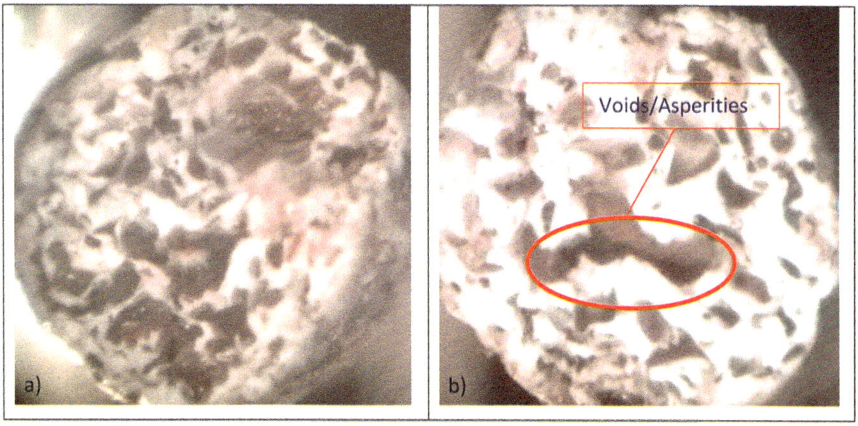

Fig. 3 Optical photomicrograph of filaments having least and highest voids

calculated S/N values for determining the predominant factor that affects the PS of the filaments.

The results of ANOVA have been shown in Table 4. From mean S/N ratio plots, it has been observed that with increase in extrusion load and temperature, PS of the composite filament increases. The increase in load helps in easy flow of the material. Also, maximum PS has been obtained at low speed, i.e. 70 rpm, because at low-speed reinforced material form sufficient strong bonds with base material which results in maximum PS. Therefore, suitable parametric conditions for obtaining maximum PS are 15 kg load, 245 °C temperature, and 70 rpm speed.

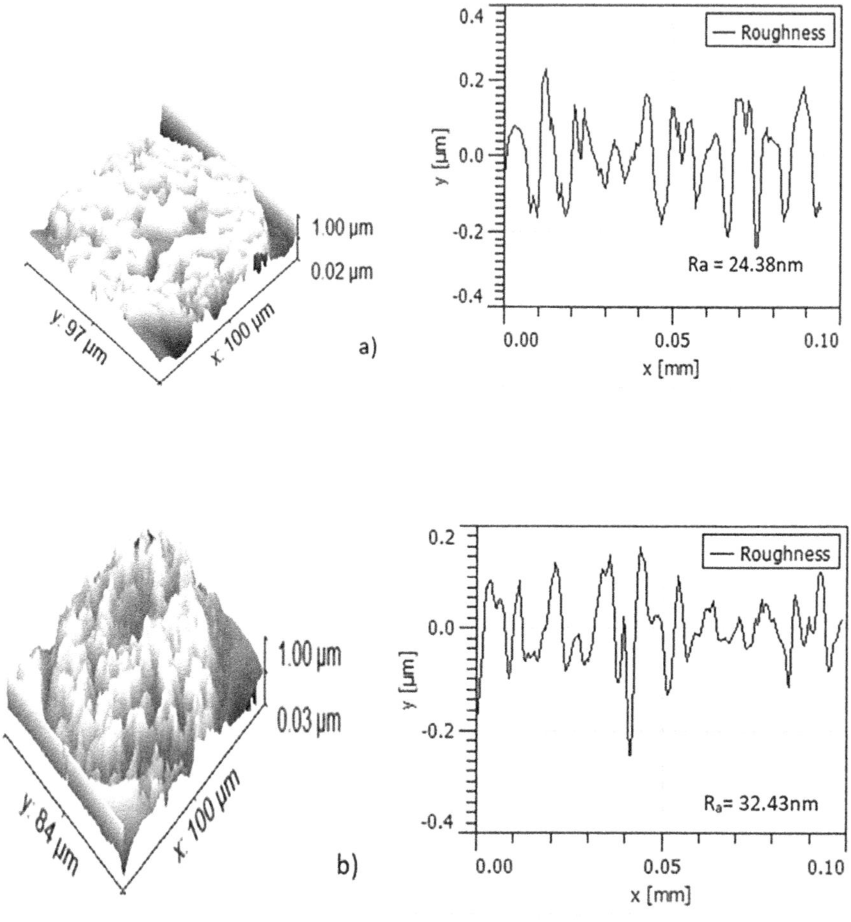

Fig. 4 Surface topology of filaments having (**a**) minimum and (**b**) maximum voids

Filaments having maximum and minimum porosity were further processed to SEM (as shown in Fig. 7) for analysing the mechanical behaviour and surface topology. From SEM analysis, it has been found that filaments having poor mechanical bonding due to improper mixing of reinforced material with base matrix results in the formation of higher number of voids. At last, thermal analysis of filaments (having maximum and minimum porosity) has been performed with the help of DSC for studying the effect of reinforcements. Thermal analysis was done at air flow of 50 ml min^{-1} using three cycles of heating and cooling for both exothermic and endothermic reactions as shown in Fig. 8. From thermal analysis, it has been observed that heat carrying capacity of sample 8 (0.86 J/g) is more as compared to sample 3 (0.63 J/g). Therefore, filament fabricated as per the processing conditions of sample 8 might be suitable for high-temperature applications.

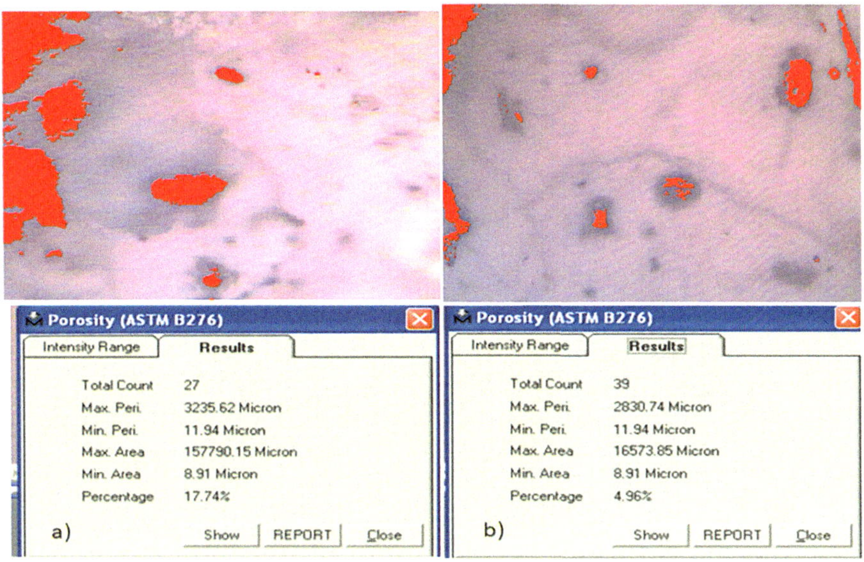

Fig. 5 Porosity presence along with percentage in composite filaments

Table 3 Calculated S/N values for PS

Sample No.	1	2	3	4	5	6	7	8	9
S/N value	19.69	19.60	19.28	21.40	20.10	22.79	23.15	26.22	25.05

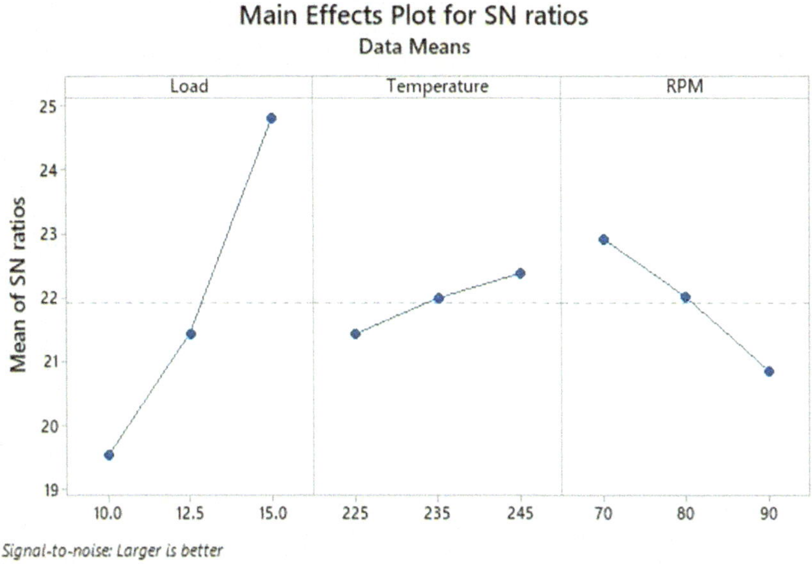

Fig. 6 Mean S/N ratio for PS of composite filaments

Table 4 ANOVA for PS of composite filaments

Source	DOF	Adj SS	Adj MS	F-value	*P*-value
Load	2	42.9665	21.4833	58.18	0.017
Temperature	2	1.3897	0.6948	1.88	0.347
Speed	2	6.3670	3.1835	8.62	0.104
Error	2	0.7386	0.3693		
Total	8	51.4617			

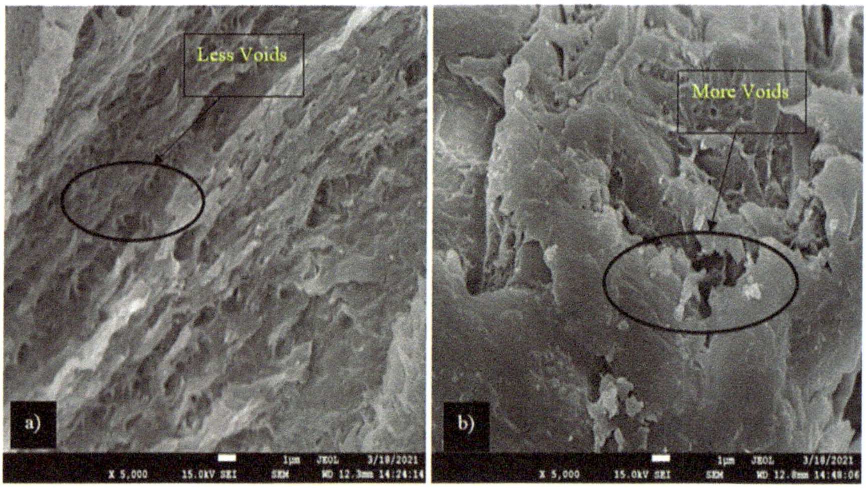

Fig. 7 SEM analysis at ×5000 for filaments having (**a**) minimum and (**b**) maximum porosity

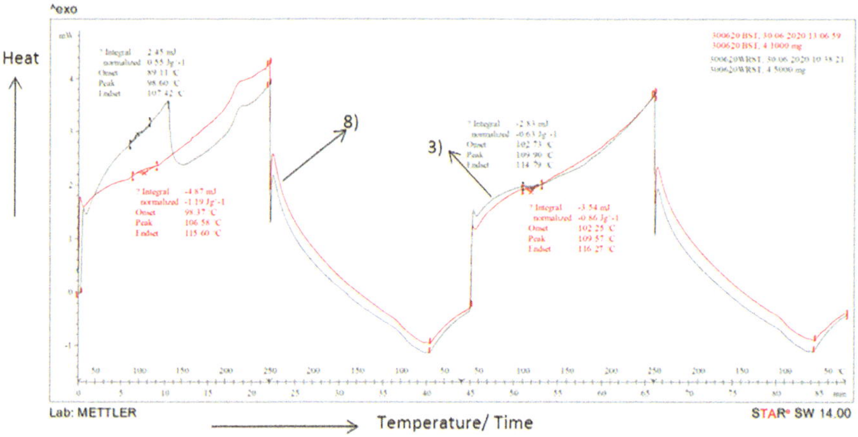

Fig. 8 Thermograms for sample 3 and sample 8

4 Conclusion

The following conclusions have been drawn from present study:

- Hybrid blending of industrial waste material in 2° recycled ABS has been successfully achieved for obtaining desired properties.
- Porosity percentage, load during extrusion and speed of extrusion affect the mechanical properties directly as filaments prepared at high speed indicated more voids and asperities which result in poor mechanical properties.
- Thermal analysis indicates that filaments containing more heat carrying capacity can be used in non-structural engineering applications. As future work, filaments with high heat carrying capacity can be used for 3D printing of non-structural engineering components.

Acknowledgements The authors are thankful to Manufacturing Research Lab GNDEC Ludhiana, Lovely Professional University and Material Characterization Lab, NITTTR Chandigarh for supporting the research work.

References

1. Chakraborty, D., Reddy, B. A., & Choudhury, A. R. (2008). Extruder path generation for curved layer fused deposition modeling. *Computer-Aided Design, 40*(2), 235–243.
2. Lee, W. C., Wei, C. C., & Chung, S. C. (2014). Development of a hybrid rapid prototyping system using low-cost fused deposition modeling and five-axis machining. *Journal of Materials Processing Technology, 214*(11), 2366–2374.
3. Conner, B. P., Manogharan, G. P., Martof, A. N., Rodomsky, L. M., Rodomsky, C. M., Jordan, D. C., & Limperos, J. W. (2014). Making sense of 3-D printing: Creating a map of additive manufacturing products and services. In *Additive manufacturing* (pp. 64–76). Elsevier.
4. Huang, Y., Leu, M. C., Mazumder, J., & Donmez, A. (2015). Additive manufacturing: current state, future potential, gaps and needs, and recommendations. *Journal of Manufacturing Science and Engineering, 137*(1), 014001.
5. Murphy, T., Gray, H., & Cotteleer, M. (2015). 3D opportunity for the future: Industry participants speak out. *Deloitte Review, 17*, 148–165.
6. Upcraft, S., & Fletcher, R. (2003). The rapid prototyping technologies. *Assembly Automation, 23*, 318–330.
7. Bakar, N. S. A., Alkahari, M. R., & Boejang, H. (2010). Analysis on fused deposition modelling performance. *Journal of Zhejiang University, Science, A, 11*(12), 972–977.
8. Greul, M., Pintat, T., & Greulich, M. (1995). Rapid prototyping of functional metallic parts. *Computers in Industry, 28*(1), 23–28.
9. Bourell, D., Stucker, B., Ilardo, R., & Williams, C. B. (2010). Design and manufacture of a Formula SAE intake system using fused deposition modeling and fiber-reinforced composite materials. *Rapid Prototyping Journal, 16*(3), 174–179.
10. Singh, R., Kumar, S., Bedi, P., & Hashmi, M. S. J. (2020). On wear of 3D printed Al_2O_3 reinforced Nylon6 matrix based functional prototypes. *Materials Today: Proceedings, 33*, 1477–1482.
11. Billah, K. M. M., Lorenzana, F. A., Martinez, N. L., Wicker, R. B., & Espalin, D. (2020). Thermomechanical characterization of short carbon fiber and short glass fiber-reinforced ABS used in large format additive manufacturing. *Additive Manufacturing, 2020*, 101299.

12. Jin, D. W., Shon, K. H., Jeong, H. M., & Kim, B. K. (1998). Compatibility enhancement of ABS/polycarbonate blends. *Journal of Applied Polymer Science, 69*(3), 533–542.
13. Lee, J. H., Shin, H., & Rhee, K. Y. (2019). Surface functionalization of boron nitride platelets via a catalytic oxidation/silanization process and thermomechanical properties of boron nitride-epoxy composites. *Composites Part B: Engineering, 157*, 276–282.
14. Pour, R. H., Soheilmoghaddam, M., Hassan, A., & Bourbigot, S. (2015). Flammability and thermal properties of polycarbonate/acrylonitrile-butadiene-styrene nanocomposites reinforced with multilayer graphene. *Polymer Degradation and Stability, 120*, 88–97.
15. Yeh, S. K., Agarwal, S., & Gupta, R. K. (2009). Wood–plastic composites formulated with virgin and recycled ABS. *Composites Science and Technology, 69*(13), 2225–2230.
16. Lohar, G. S., & Jogi, B. F. (2018). Influence of carbon black (CB) on mechanical behaviour and microscopic analysis of poly-propylene (PP)/Acrylonitrile-butadiene-styrene (ABS) nanocomposites. *Procedia Manufacturing, 20*, 85–90.
17. Love, L. J. (2014). The importance of carbon fiber to polymer additive manufacturing. *Journal of Materials Research, 29*(17), 1893–1898.
18. Chawla, K., Singh, J., & Singh, R. (2020). On recyclability of thermosetting polymer and wood dust as reinforcement in secondary recycled ABS for nonstructural engineering applications. *Journal of Thermoplastic Composite Materials*. https://doi.org/10.1177/0892705720925135
19. Panda, A. K., Singh, R. K., & Mishra, D. K. (2010). Thermolysis of waste plastics to liquid fuel: A suitable method for plastic waste management and manufacture of value added products—A world prospective. *Renewable and Sustainable Energy Reviews, 14*(1), 233–248.
20. Zhong, W., Li, F., Zhang, Z., Song, L., & Li, Z. (2001). Short fiber reinforced composites for fused deposition modeling. *Materials Science and Engineering, 301*(2), 125–133.
21. Chadha, A., Haq, M. I. U., Raina, A., Singh, R. R., Penumarti, N. B., & Bishnoi, M. S. (2019). Effect of fused deposition modelling process parameters on mechanical properties of 3D printed parts. *World Journal of Engineering, 16*, 550–559.
22. Haleem, A., & Javaid, M. (2020). 3D printed medical parts with different materials using additive manufacturing. *Clinical Epidemiology and Global Health, 8*(1), 215–223.
23. Chawla, K., Singh, J., & Singh, R. (2020). On recyclability of thermoplastic ABS polymer as fused filament for FDM technique of additive manufacturing. *World Journal of Engineering*. https://doi.org/10.1108/WJE-11-2020-0580

Mechanical and Morphological Properties Correlation of PLA-PVC-Wood Powder-Fe_3O_4 Composite Matrix for 3D Printing

Sudhir Kumar, Rupinder Singh, T. P. Singh, and Ajay Batish

1 Introduction

FDM in recent days has shown promising results in 3D printing of biocompatible polymeric composites such as PLA, polypropylene (PP), nylon-6, other polymeric composites [1, 2]. To prepare the feedstock filaments of different polymeric composites, various researchers have used different techniques such as mechanical blending, chemical blending, chemical-assisted mechanical blending, etc. [3, 4]. The changing phase of material development from 2D to 3D and now from 3D to 4D has widened the range of application of 3D printing [5, 6]. In recent days, researchers have worked on the fourth dimension of FDM which is time [7, 8]. Material that can change its shape and other characteristics with change in some stimulus which may be external or internal is called smart material matrix [9]. PLA in combination with polyvinyl chloride (PVC), wood powder and magnetite (Fe_3O_4) powder has been studied for non-structural engineering application [10]. The reinforcement of Fe_3O_4 powder in PLA matrix has resulted into improved mechanical properties [11–14] as well as magnetic characteristics in polymeric composite [15–17]. Thus, Fe_3O_4 powder alters the material behaviour and introduces the smart behaviour as material matrix becomes responsive to external magnetic field.

Reinforcement of nano graphene powder improves the mechanical properties of polymeric composite and induces piezoelectric property inside the material matrix,

S. Kumar · T. P. Singh · A. Batish
Department of mechanical Engineering, Thapar Institute of Engineering and Technology, Patiala, India

R. Singh (✉)
Department of mechanical Engineering, National Institute of Technical Teachers Training and Research, Chandigarh, India
e-mail: rupindersingh@nitttrchd.ac.in

C. Prakash et al. (eds.), *Additive, Subtractive, and Hybrid Technologies*, Mechanical Engineering Series, https://doi.org/10.1007/978-3-030-99569-0_4

thus changing the non-smart matrix into smart matrix [18–20]. Researchers have also observed that reinforcement of barium titanate (BaTiO3) and graphene powder in adequate ratio in polyvinylidene fluoride (PVDF) has resulted into enhanced piezoelectric properties [21, 22].

The literature review reveals that, in the past one decade, significant studies on mechanical and morphological properties of feedstock filament for fused deposition modelling (FDM) have been reported. But, hitherto, little has been reported on the development of correlation matrix for such properties of feedstock filament. In the present research work, an effort has been made to develop a correlation among different outputs (peak strength, break strength, melt flow index, Shore D hardness and modulus of toughness) of feedstock filament prepared with polylactic acid (PLA)-polyvinyl chloride (PVC)-wood powder-Fe_3O_4 composite matrix, which is a necessary step to predict the exact correlation among variables to thoroughly understand the mechanical performance of developed feedstock filaments.

2 Material and Methods

PLA, PVC and Fe_3O_4 powder (44 μm) were precured from Shiva Chemicals, Ludhiana. Wood powder was collected from local wood working industry. The collected wood powder was segregated to 50-μm size using sieve shaking equipment. The reinforcement of PVC was used from 10 to 25 wt%, and wood powder was used in 2.5–10 wt%. Whereas, Fe_3O_4 powder was reinforced in 10–20 wt% using mechanical blending method. Twin-screw extrusion (Model: HAAKE, German) was used to extrude the material matrix and develop feedstock filament. The prepared feedstock filament was tested for different mechanical properties using universal tensile testing machine (Make: Shanta Engineering, Pune, India, and Capacity: 5KN). The obtained mechanical properties were used to develop a correlation among different variables to understand the relationship between different variable. Table 1 shows the different variables used in the development of correlation matrix.

3 Development of Correlation Matrix and Discussions

Table 2 shows the mechanical properties of feedstock filament tested under similar condition of tensile testing using UTM machine. From Table 2, it may be observed that mechanical properties were observed to be in positive relation with shore D hardness of feedstock filaments whereas negative relationship with porosity. Whereas this observation was clear from the results, for better understanding of correlations among different variables, one must apply a statistical tool of correlation. Therefore, the data available in Table 2 was further used to develop a correlation matrix using advance excel tool.

Table 1 Different set of variables and their key information for development of correlation matrix

Variable name	Key information	Note
1. Peak strength (MPa) 2. Break strength (MPa) 3. Modulus of toughness (MPa)	Strength of material while in stretch under UTM testing	Tested for different feedstock filament materials 1. PLA/PVC (PVC 10–25 wt%) 2. PLA/wood powder (wood powder 2.5–10%) 3. PLA/Fe_3O_4 powder (Fe_3O_4 powder 10–25%)
MFI (units of MFI g/10 min)	Melt flow index of polymeric composite	
Porosity values (%)	Number of porosity holes on surface of feedstock filament seen under tool maker's microscope	
Shore D hardness (shore D)	Hardness of feedstock filament material on surface checked with shore D hardness instrument	

Table 2 Mechanical properties of developed feedstock filaments of different composites of PLA

Composition	PS	BS	MoT	MFI	Percentage porosity	Shore D hardness
Pure PLA	45.13	40.61	1.35	28.83	5.00	70.00
10% PVC	33.09	29.78	0.66	33.43	5.25	52.50
15% PVC	17.17	15.45	0.43	36.13	7.91	47.50
20% PVC	13.25	11.92	0.40	39.43	9.59	43.50
25% PVC	11.53	10.38	0.17	42.58	11.74	40.50
10% Fe_3O_4	29.17	26.25	1.17	34.56	10.56	53.50
15% Fe_3O_4	33.89	30.5	1.69	38.72	8.94	58.50
20 % Fe_3O_4	34.33	30.9	2.92	40.98	5.20	61.50
2.5% wood	41.69	37.52	0.83	25.68	6.18	58.50
5% wood	36.85	33.16	0.92	22.67	7.59	52.50
10% wood	23.41	21.07	0.59	18.53	9.75	48.50

PS peak strength, *BS* break strength, *MoT* modulus of toughness

Figure 1 shows the (a) correlation matrix of peak strength with MFI, porosity and shore D hardness as other variables and (b) correlation graph among different variables.

From correlation matrix of peak strength, it may be observed that, peak strength and shore D hardness have strong positive relationship as their correlation value came out to be 0.91 which is very near to 1. Whereas, there is strong negative correlation with melt flow index and peak strength (−0.44). Porosity and shore D hardness have shown negative correlation of −0.73 as well as porosity with peak strength has shown negative correlation of −0.74 with peak strength. The correlation matrix signifies that with increase in hardness peak strength increases, porosity decreases. Whereas, melt flow index has very poor correlation with other variables. Similar understanding may be developed from its correlation graph. From correlation matrix and graph for break strength property as observed from Fig. 2a, b, it may be observed that the correlation matrix and coefficients were very similar to the matrix of

Properties	Peak Strength	Melt Flow Index	Percentage Porosity	Shore D Hardness
Peak Strength	1.00	-0.44	-0.74	0.91
Melt Flow Index	-0.44	1.00	0.20	-0.21
Percentage Porosity	-0.74	0.20	1.00	-0.73
Shore D Hardness	0.91	-0.21	-0.73	1.00

(a)

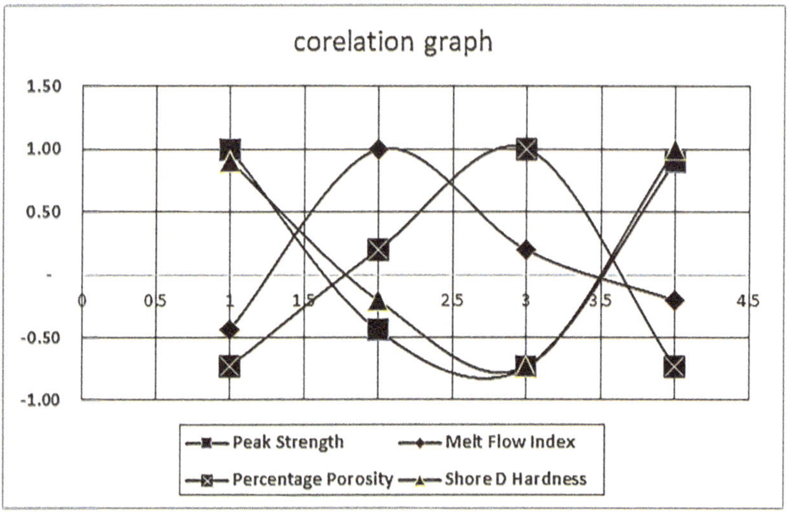

(b)

Fig. 1 (**a**) A 4 × 4 corelation matrix where columns 1, 2, 3 and 4 represent peak strength, melt flow, percentage prosody and shore D hardness, respectively, and rows 1, 2, 3 and 4 represent the peak strength, melt flow, percentage prosody and shore D hardness, respectively. Corelation matrix values varies between −1 and 1, depending over the corelation of properties with each other. (**b**) The corelation graph for the same data

Properties	Break Strength	Melt Flow Index	Percentage Porosity	Shore D Hardness
Beak Strength	1.00	-0.44	-0.74	0.91
Melt Flow Index	-0.44	1.00	0.20	-0.21
Percentage Porosity	-0.74	0.20	1.00	-0.73
Shore D Hardness	0.91	-0.21	-0.73	1.00

(a)

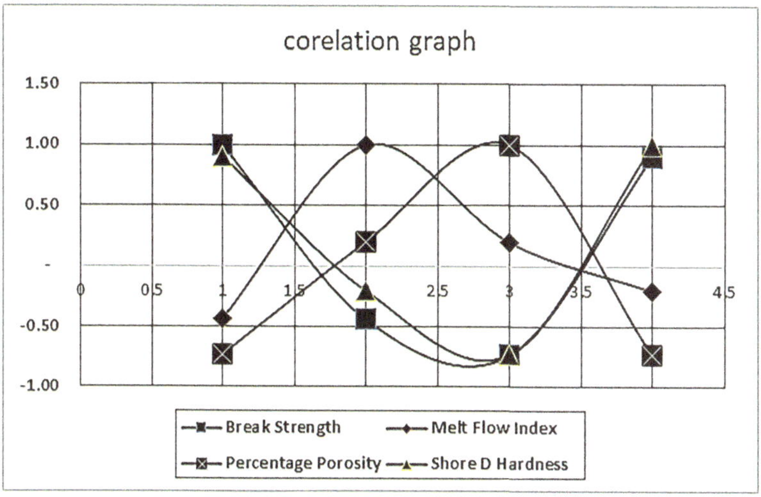

(b)

Fig. 2 (**a**) A 4 × 4 corelation matrix where columns 1, 2, 3 and 4 represent break strength, melt flow, percentage prosody and shore D hardness, respectively, and rows 1, 2, 3 and 4 represent the break strength, melt flow, percentage prosody and shore D hardness, respectively. Corelation matrix values varies between -1 and 1 depending over the corelation of properties with each other. (**b**) The corelation graph for the same data

Properties	Modulus of toughness	Melt Flow Index	Percentage Porosity	Shore D Hardness
Modulus of toughness	1.00	0.20	-0.49	0.71
Melt Flow Index	0.20	1.00	0.20	-0.21
Percentage Porosity	-0.49	0.20	1.00	-0.73
Shore D Hardness	0.71	-0.21	-0.73	1.00

(a)

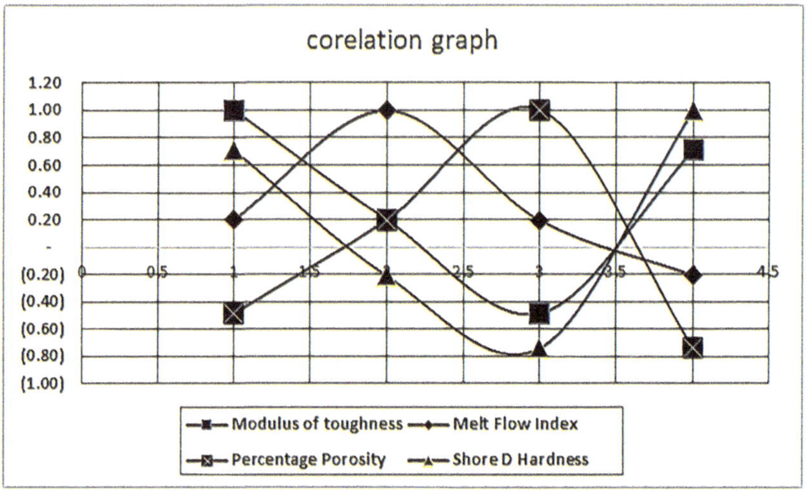

(b)

Fig. 3 (**a**) A 4 × 4 corelation matrix where columns 1, 2, 3 and 4 represent modulus of toughness, melt flow, percentage prosody and shore D hardness, respectively, and rows 1, 2, 3 and 4 represent the modulus of toughness, melt flow, percentage prosody and shore D hardness, respectively. Corelation matrix values vary between −1 and 1 depending on the corelation of properties with each other. (**b**) The corelation graph for the same data

correlation for peak strength. Thus peak strength and break strength were observed to be in strong positive relation with each other.

From correlation matrix and correlation graph for modulus of toughness (see Fig. 3a, b), it was observed that shore D hardness was in positive relationship with modulus of toughness, whereas porosity held negative relationship. Thus with increasing porosity, there was a decrease in modulus of toughness which may be observed from the table. Whereas, melt flow index has shown poor relationship with any other parameter, thus showing that the melt flow index is result of internal characteristics of material composition. Shore D hardness and porosity have shown negative correlation of −0.73. Thus, specifying that with increase in the porosity level, the surface hardness decreases significantly.

4 Conclusions

The present work developed different correlation matrix among different variables of feedstock filament development on lab scale, and from the study, following points may be concluded:

1. Melt flow index of material has shown least correlation with other variables, thus it may be concluded that melt flow index is an inherent property of material and is least bothered with other processing variables.
2. Shore D hardness has strong positive correlation with mechanical properties of feedstock filament and negative correlation with porosity of material matrix. Thus, to have better mechanical properties, the feedstock filament must have high surface hardness and low porosity.
3. Porosity has shown strong negative correlation of −0.73 and −0.74 with shore D hardness and peak strength, respectively.

Acknowledgement The authors are highly obliged to Thapar Institute of Engineering and Technology, Patiala and NITTTR Chandigarh for technical support.

References

1. Cuiffo, M. A., Snyder, J., Elliott, A. M., Romero, N., Kannan, S., & Halada, G. P. (2017). Impact of the fused deposition (FDM) printing process on polylactic acid (PLA) chemistry and structure. *Applied Sciences, 7*(6), 579.
2. Heidari-Rarani, M., Rafiee-Afarani, M., & Zahedi, A. M. (2019). Mechanical characterization of FDM 3D printing of continuous carbon fiber reinforced PLA composites. *Composites Part B: Engineering, 175*, 107147.
3. Beesetty, P., Patil, B., & Doddamani, M. (2020). Mechanical behavior of additively manufactured nanoclay/HDPE nanocomposites. *Composite Structures, 247*, 112442.
4. Sandhu, G. S., & Singh, R. (2019). Development of ABS-graphene blended feedstock filament for FDM process. In *Additive manufacturing of emerging materials* (pp. 279–297). Springer.

5. Kumar, S., Singh, R., Singh, T. P., & Batish, A. (2019). On investigation of rheological, mechanical and morphological characteristics of waste polymer-based feedstock filament for 3D printing applications. *Journal of Thermoplastic Composite Materials*. https://doi.org/10.1177/0892705719856063
6. Momeni, F., Liu, X., & Ni, J. (2017). A review of 4D printing. *Materials & Design, 122*, 42–79.
7. Kuang, X., Roach, D. J., Wu, J., Hamel, C. M., Ding, Z., Wang, T., Dunn, M. L., & Qi, H. J. (2019). Advances in 4D printing: Materials and applications. *Advanced Functional Materials, 29*(2), 1805290.
8. Kumar, S., Singh, R., Batish, A., & Singh, T. P. (2019). Additive manufacturing of smart materials exhibiting 4-D properties: A state of art review. *Journal of Thermoplastic Composite Materials*. https://doi.org/10.1177/0892705719895052
9. Kumar, S., Singh, R., Singh, T. P., & Batish, A. (2020). Fused filament fabrication: A comprehensive review. *Journal of Thermoplastic Composite Materials*. https://doi.org/10.1177/0892705720970629
10. Miao, S., Castro, N., Nowicki, M., Xia, L., Cui, H., Zhou, X., Zhu, W., Lee, S. J., Sarkar, K., Vozzi, G., & Tabata, Y. (2017). 4D printing of polymeric materials for tissue and organ regeneration. *Materials Today, 20*(10), 577–591.
11. Kumar, S., Singh, R., Singh, T. P., & Batish, A. (2019). On mechanical characterization of 3-D printed PLA-PVC-wood dust-Fe3O4 composite. *Journal of Thermoplastic Composite Materials*. https://doi.org/10.1177/0892705719879195
12. Kumar, S., Singh, R., Singh, T. P., & Batish, A. (2019). Investigations for magnetic properties of PLA-PVC-Fe3O4-wood dust blend for self-assembly applications. *Journal of Thermoplastic Composite Materials*. https://doi.org/10.1177/0892705719857778
13. Kumar, S., Singh, R., Singh, T. P., & Batish, A. (2020). Investigations for mechanical, thermal and magnetic properties of polymeric composite matrix for four-dimensional printing applications. *Journal of the Brazilian Society of Mechanical Sciences and Engineering, 42*(4), 1–15.
14. Raksaksri, L., Chaiwutthinan, P., Larpkasemsuk, A., Chuayjuljit, S., & Boonmahitthisud, A. (2020). Properties of wood-plastic composites based on pvc/pla/pbat ternary blend. *Journal of Metals, Materials and Minerals, 30*, 2.
15. Kumar, S., Singh, R., Singh, T. P., & Batish, A. (2020). Flexural, pull-out, and fractured surface characterization for multi-material 3D printed functionally graded prototype. *Journal of Composite Materials, 54*(16), 2087–2099.
16. Kumar, S., Singh, R., Singh, M., Singh, T. P., & Batish, A. (2020). Multi material 3D printing of PLA-PA6/TiO2 polymeric matrix: Flexural, wear and morphological properties. *Journal of Thermoplastic Composite Materials*. https://doi.org/10.1177/0892705720953193
17. Kumar, S., Singh, R., & Hashmi, M. S. J. (2020). Metal matrix composite: A methodological review. *Advances in Materials and Processing Technologies, 6*(1), 13–24.
18. Camargo, J. C., Machado, Á. R., Almeida, E. C., & Silva, E. F. M. S. (2019). Mechanical properties of PLA-graphene filament for FDM 3D printing. *The International Journal of Advanced Manufacturing Technology, 103*(5), 2423–2443.
19. Ivanov, E., Kotsilkova, R., Xia, H., Chen, Y., Donato, R. K., Donato, K., Godoy, A. P., Di Maio, R., Silvestre, C., Cimmino, S., & Angelov, V. (2019). PLA/graphene/MWCNT composites with improved electrical and thermal properties suitable for FDM 3D printing applications. *Applied Sciences, 9*(6), 1209.
20. Yao, T., Ye, J., Deng, Z., Zhang, K., Ma, Y., & Ouyang, H. (2020). Tensile failure strength and separation angle of FDM 3D printing PLA material: Experimental and theoretical analyses. *Composites Part B: Engineering, 188*, 107894.
21. Kim, H., Fernando, T., Li, M., Lin, Y., & Tseng, T. L. B. (2018). Fabrication and characterization of 3D printed BaTiO3/PVDF nanocomposites. *Journal of Composite Materials, 52*(2), 197–206.
22. Sharma, R., Singh, R., & Batish, A. (2020). On effect of chemical-assisted mechanical blending of barium titanate and graphene in PVDF for 3D printing applications. *Journal of Thermoplastic Composite Materials*. https://doi.org/10.1177/0892705720945377

Analysis of Machining Parameters in Drilling of Biocompatible Composite: HAp-HDPE and HAp-UHMWPE

Arpan Mondal, Suman Chatterjee, Anshuman Kumar Sahu, Siba Sankar Mahapatra, and Chander Prakash

1 Introduction

Composite materials are basically materials made from more than two constituent materials having diverse physical or chemical properties. When the materials are combined, it produces a material with different properties from the individual constituent materials. Each individual component remains separate and different within the finished composite material. Generally, composite materials consisting building materials of cements, concrete, reinforced plastics like fiber-reinforced polymer, metal composites, and ceramic composites. A bio-composite is a material made by a matrix (resin) and a reinforcement of natural fibers (usually derived from plants or cellulose). Orthopedic problems like wear in the joints, joint fracture, and bone defects are common these days due to heavy work load. Therefore, an alternative for grafting or implantation by orthopedic surgery is essentially required. Bone is said to be an ecclesiastical shape formed by constituents like calcium, carbonated hydroxyapatite, type I collagen, non-collagenous proteins, and water [1]. Mineral contents in bone tissue lie in the range of 50–70% [2, 3]. Flexibility, rigidity, and nutrient contents of bones depend upon collagen, minerals, and water present in the bone, respectively. Hydroxyapatite (HAp) [$Ca_{10}(PO_4)_6(OH)_2$] helps in constructing

A. Mondal
Department of Mechanical Engineering, National Institute of Technical Teachers Training and Research, Kolkata, India

S. Chatterjee · A. K. Sahu (✉) · S. S. Mahapatra
Department of Mechanical Engineering, National Institute of Technology Rourkela, Rourkela, Odisha, India

C. Prakash
Department of Industrial Engineering, Lovely Professional University, Phagwara, Punjab, India

C. Prakash et al. (eds.), *Additive, Subtractive, and Hybrid Technologies*, Mechanical Engineering Series, https://doi.org/10.1007/978-3-030-99569-0_5

the bone tissue [4]. HAp reinforced with polymers provides an option to generate synthetic bone with tailored mechanical, biological, and surgical functions [5]. Composites with reinforcement of Hap in polymers used in the recent period are HAp/ZrO_2, HAp/methacrylate, HAp/TCP (tricalcium phosphate), HAp/HDPE (high-density polyethylene), ultra-high molecular weight polyethylene (UHMWPE) composites, etc. to develop materials as bone substitute materials [6–10].

In many ways, these bones are grafted, Masatakadeie et al. [11] suggested artificial bone grafting and core decompression for continuous osteonecrosis of the femoral condyle in the knee. Interconnected ceramic hydroxyapatite in surgical treatment of bone gives an option for curettage of benign bone which gives an option in surgical uses [12–15]. Grafting of bones is done on the part of damage portion of the bone. Surgeons usually do grafting by drilling. The implant bones are needed to be secured by screw and nut. When drilling is carried out, important parameters of drilling need to be controlled, so that minimum taper and maximum circularity at entry and exit can be achieved. Machining parameters affecting drilling of bio-composites are HAp volume percentage, speed, feed, and drill diameter. Experiments are carried out as per Taguchi's L_{16} orthogonal array with mixed-level design, varying the factors at different levels on a CNC MAXMILL machine (MTAB, India). Taguchi's design of experiment approach is used to obtain maximum information about the process with less number of experiments [16–18]. Tsao and Hocheng [19] have used Taguchi method for parametric analysis of thrust of force of drill. Erol [20] suggests that Taguchi method is a powerful tool for the optimization of cutting parameters. However, Taguchi method is suitable for the optimization of a single response. When multiple responses are dealt simultaneously, the method breaks down. Therefore, many methods have been suggested to deal with multiple responses. For example, Huang and Lin [21] suggested a regression and principal component analysis (PCA)-based method to obtain optimum injection molding parameters in a multiple quality characteristics situation. Tong et al. [22] have suggested that optimization of multi-responses by principal component analysis leads to superior result as the method takes into account correlation of responses. The responses for this work are circularity at entry, circularity at exit, and taper of the holes, and they are suspected to be correlated. Principal component analysis, a data reduction technique, is capable of expressing the correlated responses into uncorrelated principal components [22–27]. The percentage of variation explained by each component is considered as the weight for the component. A linear combination of weights and components is capable of expressing multiple responses in a single equivalent response.

2 Experimental Analysis

2.1 Preparation of Composites and Characterization

Hydroxyapatite (HAp) [$Ca_{10}(PO_4)_6(OH)_2$] powder is prepared by wet chemical precipitation route method by mixing calcium hydroxide ($Ca(OH)_2$) powder, orthophosphoric (H_3PO_4) acid, and ammonia (NH_3) solution [4, 28–32]. HAp is reinforced with polymer to produce the composite materials which can possibly be implanted as artificial bone. In preparing HAp, first 10 g of calcium hydroxide is weighted in a weighing machine (Mettler Toledo India, 0.01 g accuracy). Water of weight nearly 40 times the weight of HAp is mixed to form the solution. The solution is mixed by a magnetic stirrer (Remi Equipments Pvt. Ltd., India) at 50 °C to 60 °C for 3–4 h. Then, ortho-phosphoric acid is mixed with the solution with a rate of 30 drops per minute by a burette and constantly stirred without heat input to the solution until the pH value reach to 8–10. When the pH value of the solution reduces below the threshold values, then ammonia solution is added to increase the pH value. After 5–6 h, the magnetic stirrer is stopped, and the mixer is kept for 12 h at room temperature. Then, precipitations of mixture are collected using a filter paper. HAp precipitations are dried at 80 °C in an oven and calcinated at 850 °C for 2–3 h in a muffle furnace and finally, powder HAp is obtained.

Sintered HAp powder and commercial grades of high-density polyethylene (HDPE) and ultra-high molecular weight polyethylene (UHMWPE) are mixed to fabricate composites (HAp-HDPE and HAp-UHMWPE) by micro-injection molding. Composites with different amounts of HAp (10, 20, 30, 40 vol%) are produced by mixing, melting, granulating, and injection molding. First, spray-dried HAp powder (density 3.154 g/cm^3 estimated by water immersion method and mean particle size of 19.94 μm) is mixed with polymer matrix with a DSM XPLORE micro-compounder. The mixtures are then melted, granulated, and dried at 160 °C for 2 h prior to the next processing stage [33]. Then test specimens are produced by injection molding [34]. The micro-injection molding barrel temperature at different zones are 180 °C, 190 °C, 200 °C, 220 °C for zone 1, zone 2, zone 3, zone 4, respectively, for HAp-HDPE composite. The mold temperature is 60 °C and injection pressure of 14 MPa. The time spent for loading, holding, and cooling of HDPE is 4 s, 30 s, and 30 s, respectively.

Melt flow index of UHMWPE is very low (0.09 gm/10 min at 190 °C); hence, micro-injection molding is difficult due to fail of screw with high load. Therefore, compression molding route has been adopted. Sintered HAp powder and commercial grade UHMWPE is mixed in a Thermo Scientific (TSE 24 MC) twin screw extruder at a temperature between 180 °C and 220 °C. The extruded material is pelletized into granules and then twin screw extruded at 30 rpm. After granulation, compression molding technique is applied to get the composite sheets. The specimens are hot drawn at 100 °C having a cross-head speed of 50 mm/min. The granules are vacuum dried at 80 °C overnight to remove any traces solvent. A NEOPLAST hydraulic press (Model no-HP 80T) is used at operating temperature of 220 °C in both the top and the bottom plates and applying a pressure of 308 MPa. A contour cutter is used to cut the samples into required shape.

The XRD graph of HAp-HDPE composite is shown in Fig. 1 for different volume percentage of HAp. The high peaks show dry HAp in its pure form with reference code 74-0566.

Micro-structural examination is done by using JSM JEOL 6480 LV Scanning Electron Microscope (SEM) for both the composites. Micrographs indicate that

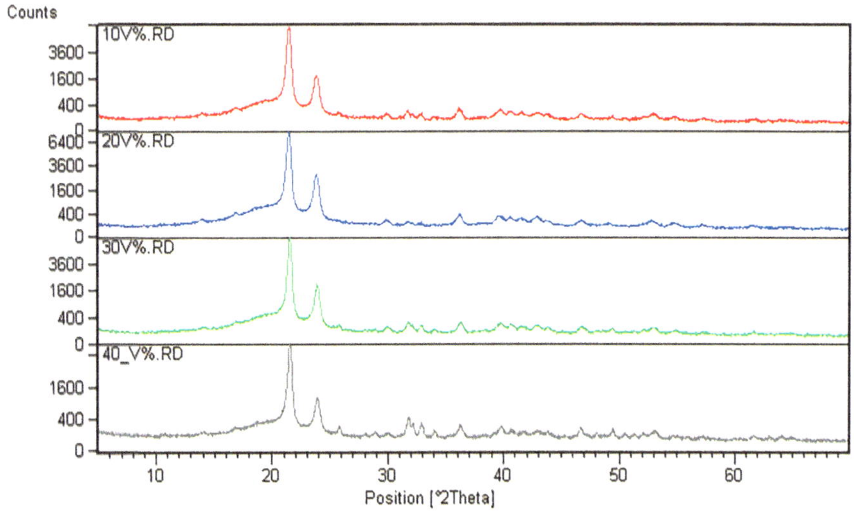

Fig. 1 XRD graph for different vol. % of HAp with HDPE

HAp is uniformly distributed over the matrix. However, some globules are observed due to in homogenous mixing (Figs. 2 and 3).

The composites were prepared in standard laboratory atmosphere and ASTM standards are followed for preparing the specimens to perform different mechanical tests. Tensile, compressive, and flexural tests are carried out in INSTRON 3382 UTM machine. Impact test is carried out in Tinius Olsen Impact Systems IT406. The specifications for all tests are listed in Table 1.

The graph between Young's modulus versus HAp vol. % is shown in Fig. 4. It can be observed that both the composites exhibit maximum Young's modulus for 40 vol. % of HAp. However, HAp-UHMWPE composite shows higher Young's modulus for any volume percentage as compared to HAp-HDPE composite.

Figures 5, 6, and 7 show the graph between HAp vol. % with tensile strength, HAp vol. % versus compressive strength, HAp vol. % versus flexural strength, and HAp vol. % versus inter-laminar shear stress (ILSS), respectively. Figure 5 shows that tensile strength is decreasing with the increasing amount of HAp vol. % for both HAp-HDPE and HAp-UHMWPE composites. However, reduction of tensile strength with increase of HAp volume percentage is more pronounced for HAp-UHMWPE composite. Figure 6 shows that when the HAp vol. % is increased, compressive strength is decreased for HAp-UHMWPE composite. However, decrease of compressive strength with the increase of HAp vol. % is not appreciable in case of HAp-HDPE composite.

Flexural test is done by an INSTRON 3382 UTM as per ASTM D790 standard at a crosshead speed of 0.77 mm/min. The dimension of sample specimen is 3.2 × 12.7 × 165 mm. The flexural strength (FS) of the specimen is calculated as follows:

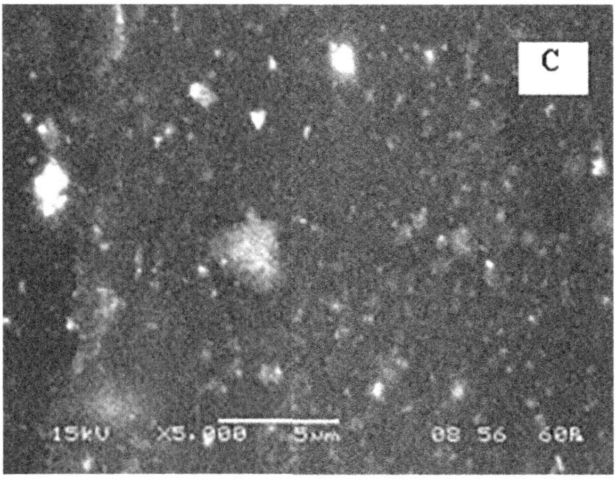

Fig. 2 SEM image of HAp-HDPE composite at 30 vol. % of HAp

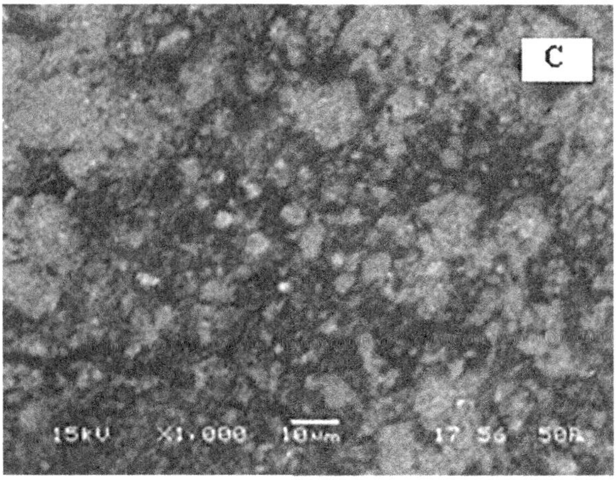

Fig. 3 SEM image of HAp-UHMWPE at 30 vol. % of HAp

Table 1 Specifications of all the testing machines

	Machine used	Variables	Standard used
Tensile	INSTRON 3382 UTM	Load cell: 100 KN Rate: 10 mm/min	ASTM D638
Compressive	INSTRON 3382 UTM	Rate: 1.3 mm/min	ASTM D695
Flexural	INSTRON 3382 UTM	Rate: 0.77 mm/min	ASTM D790
Impact	Tinius Olsen Impact Systems IT406	–	ASTM D256

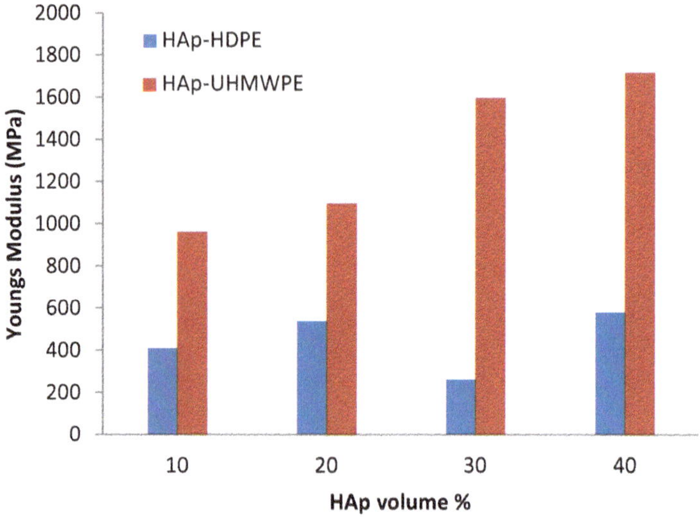

Fig. 4 Plot of Young's modulus versus HAp vol. %

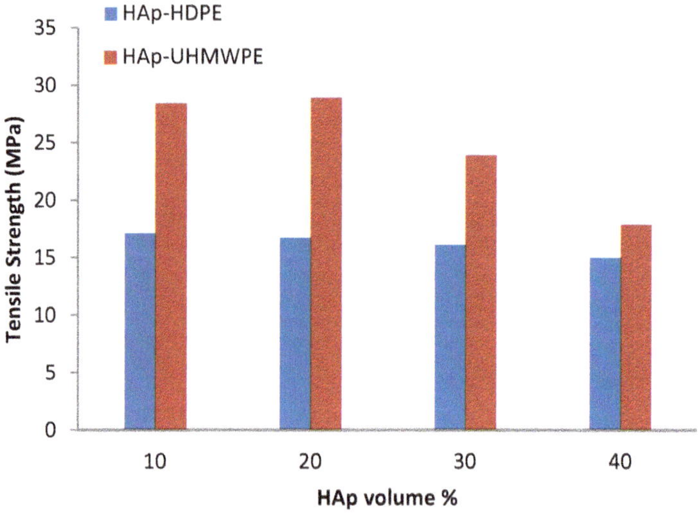

Fig. 5 Effects of HAp contents on the tensile strength

$$FS = \frac{3P.L}{b.t^3} \tag{1}$$

where P is the maximum load, b is the width, t is the thickness, and L is the length of the specimen. The data recorded in the flexural test is also used to determine the inter-laminar shear stress (ILSS)) values calculated as follows:

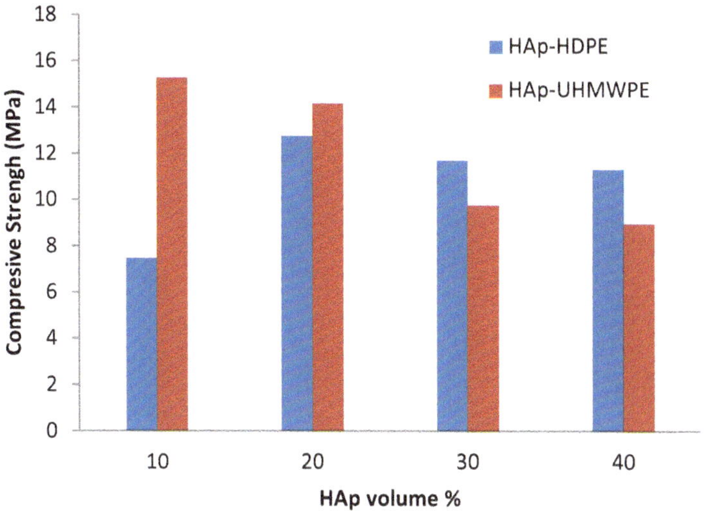

Fig. 6 Effects of HAp contents on the compressive strength

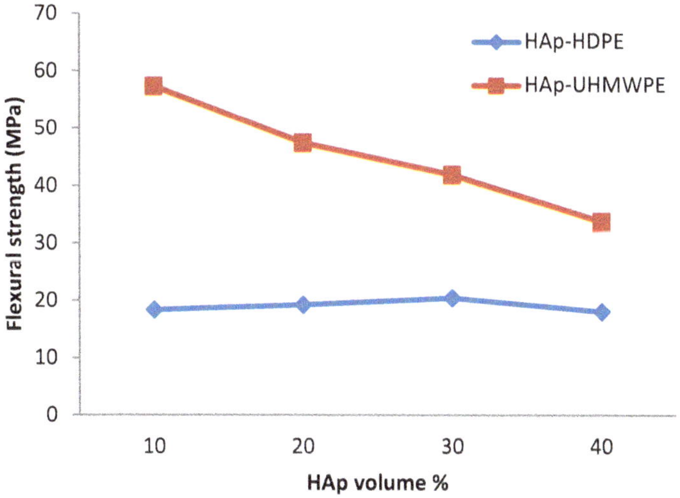

Fig. 7 Effects of HAp contents on the flexural strength

$$\mathrm{LSS} = \frac{3P}{4b.t} \tag{2}$$

The flexural properties are of great importance for any structural element. Biocompatible composite used in hip joints may fail in bending loads, and therefore, the development of new composites with improved flexural characteristics is desirable. Figure 7 shows that flexural strength of HAp-UHMWPE composite is

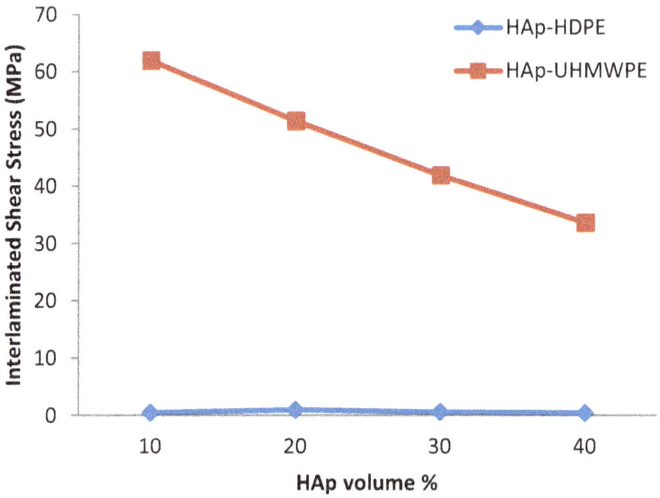

Fig. 8 Reinforcement effect of HAp on ILSS)

higher than HAp-HDPE composite for any value of HAp content. However, flexural strength decreases with increase of HAp volume % for HAp-UHMWPE composite. No appreciable change in flexural strength is noticed for HAp-HDPE composite with increase in HAp content. Figure 8 shows the plot between HAp vol. % versus ILSS) in which 20 vol. % of HAp gives maximum ILSS) for HAp-HDPE, but the plot between HAp vol. % versus ILSS for HAp-UHMWPE composite shows decrease of ILLS with increase in HAp content.

The tensile properties of HAp-HDPE and HAp-HMHDPE composites at 20% vol. of HAp are compared with previous studies. The present study gives a tensile strength of 17.18 MPa, whereas past studies report tensile strength as 17.65–19.97 MPa [35–37] for HAp-HDPE composites. The strength values reported in the present study is comparable with past studies. It is observed that the present study gives a tensile strength of 28.21 MPa, whereas Fang et al. [38] obtained the tensile strength of 26.67 MPa.

2.2 *Drilling of Composites*

Drilling is a cutting process that uses a drill bit to cut or enlarge a hole of circular cross-section in solid materials. The composites under study being softer than most metals, drilling of composite is considerably easier and faster. Cutting fluids are not used or needed. Here, the effect of different parameter settings on circularity at entry and exit and hole taper on HAp-HDPE composite is analyzed. The thickness

Table 2 Drilling parameters and with levels

Machining parameters	Level 1	Level 2	Level 3	Level 4
HAp (vol. %)	10	20	30	40
Feed rate (mm/min)	50	100	150	200
Speed (rpm)	750	1500	2250	3000
Drill diameter (mm)	5	6		

Table 3 Taguchi's mixed-level L_{16} orthogonal array combinations

Experimental runs	HAp (vol. %)	Feed rate (mm/min)	Speed (rpm)	Drill diameter (mm)
1	10	50	750	5
2	10	100	1500	5
3	10	150	2250	6
4	10	200	3000	6
5	20	50	1500	6
6	20	100	750	6
7	20	150	3000	5
8	20	200	2250	5
9	30	50	2250	5
10	30	100	3000	5
11	30	150	750	6
12	30	200	1500	6
13	40	50	3000	6
14	40	100	2250	6
15	40	150	1500	5
16	40	200	750	5

of the composite sample is measured at different sections with a digital vernier caliper with least count of 0.01 mm. Then, drilling process are done on the composite work piece at different parameter settings. The diameters of the drilled holes are in millimeter range, and the holes are drilled without any relative motion between the job and the workpiece. The experiments were done on a CNC MAXMILL machine system. The drilling operation is done on HAp-HDPE composite work piece with a mean thickness of 3.03 mm, 2.95 mm, 3.12 mm, and 3.18 mm for 10 vol. %, 20 vol. %, 30 vol. %, and 40 vol. %, respectively. The controllable parameters considered with their levels are shown in Table 2. Experiments are carried out using Taguchi's L_{16} orthogonal array. The factorial combinations for experimental runs are shown in Table 3. The drilling operation is shown in Fig. 9.

Fig. 9 Drilling operation

3 Multi-response Optimization Using Principal Component Analysis

Principal component analysis is a projection-based technique that facilitates reduction in data dimension through the construction of orthogonal principal components that are weighted linear combinations of the original variables. To optimize the multiple responses into a single equivalent response, various methods are used. However, the responses may be correlated among themselves. Therefore, the responses must be uncorrelated before finding an equivalent response. Tong et al. [22] have used principal component analysis along with technique for order preference by similarity to ideal solution (TOPSIS) for multiple response optimization of processing thin films to be used in integrated circuit [22]. Liu et al. introduces PCA algorithm for network wide traffic anomaly detection in a distributed fashion [39]. Here, the effect of important drilling parameters like HAp vol. %, feed, speed, and drill diameter on the responses such as taper, circularity at entry, and circularity at exit are considered. The multi-responses are converted into a single response using principal component analysis (PCA); hence, the influence of correlation among the responses can be eliminated. Steps including for the procedure are as follows:

Step 1. Normalize the responses [21] by using the Eqs. (3) and (4) for larger the better and smaller the criterion, respectively. Since taper is to be minimized, smaller-the-better criterion is to be used. As the circularity at entry and exit is expressed as the ratio of minimum Farret diameter to maximum Farret diameter, it is maximized using larger-the-better characteristic is used.

$$\text{Larger the better}, Y_{ij} = \frac{L_{ij} - \min(L_j)}{\max(L_j) - \min(L_j)} \tag{3}$$

$$\text{Smaller the better}, Y_{ij} = \frac{\max(L_j) - L_{ij}}{\max(L_j) - \min(L_j)} \tag{4}$$

Step 2. Calculate the PCA on the normalized values of the responses to obtain principal component scores (PCSs) [36], to find the eigen values and vectors for the corresponding responses by getting the relation PCS_{ij}, computing $(g)_{m\times p}$ to PCA, the value of ith ($i = 1, 2, \ldots, p$) PCS corresponding to ith trial, (PCS_{il}) can be obtained as follows:

$$PCS_{il} = a_{k1}n_{i1} + a_{k2}n_{i2} + \ldots + a_{kp}n_{ip} \tag{5}$$

where a_{k1}, a_{k2},...,a_{kp} are the eigen vectors with respect to eigenvalues (kth terms) of the matrix.

Step 3. Next we have to perform normalization of the PCSs (X_{il}) values [39], by the following equation:

$$X_{il} = \frac{PCS_{il} - \min(PCS_l)}{\max(PCS_l) - \min(PCS_l)} \tag{6}$$

Step 4. Three principal components are obtained using PCA, that principal component 1 (PC1) has V_1 of variance, (PC2) has V_2, and (PC3) has V_3 of total variance. The relation for weighted principal component, WPC, is given by equation:

$$WPC = (V_1 \times PCS1) + (V_2 \times PCS3) + (V_3 \times PCS3) \tag{7}$$

4 Results and Discussions

The effect of different drilling parameters on circularity at entry and exit and hole taper on HAp-HDPE bio-composite is analyzed. The circularity at both entry and exit is calculated by using the ratio of minimum to maximum diameters of the hole. After measuring the entrance diameter and exit diameter of the hole, the hole taper is calculated by Eq. (8).

$$\text{Hole taper (in radian)} = \frac{(\text{hole enterance diameter}) - (\text{hole exit diameter})}{(2 \times \text{thickness of the workpiece})} \tag{8}$$

The output responses are calculated for taper, circularity at entry, and circularity at exit as shown in Table 4.

Table 4 Design of L_{16} experiments and its observations

Sl. No.	HAp (vol. %)	Feed rate (mm/min)	Speed (rpm)	Drill diameter (mm)	Taper (radian)	Circularity at entry (mm)	Circularity at exit (mm)
1	10	50	750	5	0.0242	0.9642	0.8217
2	10	100	1500	5	0.0098	0.8415	0.7821
3	10	150	2250	6	0.0018	0.8495	0.8388
4	10	200	3000	6	0.0020	0.8666	0.8544
5	20	50	1500	6	0.0046	0.8356	0.8084
6	20	100	750	6	0.0017	0.845	0.835
7	20	150	3000	5	0.0006	0.8352	0.8314
8	20	200	2250	5	0.0009	0.8554	0.8502
9	30	50	2250	5	0.0056	0.8514	0.8166
10	30	100	3000	5	0.0040	0.8379	0.8128
11	30	150	750	6	0.0003	0.8402	0.8383
12	30	200	1500	6	0.0009	0.8374	0.8316
13	40	50	3000	6	0.0037	0.8238	0.8000
14	40	100	2250	6	0.0023	0.8272	0.8125
15	40	150	1500	5	0.0008	0.8238	0.8187
16	40	200	750	5	0.0019	0.8474	0.8352

Table 5 Eigenvalues, eigenvectors, accountability proportion (AP), and cumulative variation (CV) computed for the first three major quality indicators

Principal component	Eigen value	Explained variation	Cumulative variation (%)	Eigen vector (taper, circularity at entry, circularity at exit)
PC1	1.8701	0.623	0.623	[0.723, −0.671, 0.163]
PC2	1.1298	0.377	1	[−0.140, −0.374, −0.917]
PC3	0.0001	0.000	1	[−0.676, −0.640, 0.364]

4.1 Principal Component Analysis

Optimization of the observed data has been done in Table 4, by using the responses and normalized by the Eqs. (1) and (2) and formulated in Minitab 16 which gives the eigen values and vectors tabulated in Table 5.

The PCS values were obtained using Eq. (3) and normalized by using Eq. (4). This is further optimized by using Eq. (5) tabulated in Table 4 to get the weighted WPC. Tables 6 and 7 shows the optimum condition obtained using Minitab 16. An L_{16} array is performed to get the analysis of variance (ANOVA) and signal-to-noise ratio shown in Tables 8 and 9, for the study of machining parameters showing the s/n ratio plot of WPC in Fig. 10.

The main effects plot for means of SN ratios of WPC as in Fig. 10. Here, higher-the-better (HTB) value is considered, and it shows the optimum parametric setting as $A_1B_1C_2D_1$. That is, feed rate of 50 mm/min, speed of 1500 rpm, and 5 mm drill diameter for drill in 10 vol. % of HAp sample, and then we get minimum taper with maximum circularity at both entry and exit [40–45].

Table 6 Principal component analysis for L_{16} experimental observations

Sl. No.	Normalization of outputs Y_{ij}			Principal components		
	Y_{1j}	Y_{2j}	Y_{3j}	PCS1	PCS2	PCS3
1	6.56×10^{-6}	0	−0.45228	−0.0737	0.414742	−0.16464
2	0.60583	−0.87393	−1	0.86144	1.15903	−0.21424
3	0.96091	−0.81695	−0.21577	1.20774	0.368871	−0.20527
4	0.94997	−0.69516	0	1.15328	0.126992	−0.19728
5	0.83524	−0.91595	−0.63624	1.11477	0.809063	−0.21000
6	0.96404	−0.849	−0.26833	1.22294	0.428617	−0.20600
7	1.01047	−0.9188	−0.31812	1.29523	0.493882	−0.21084
8	0.99998	−0.77493	−0.05809	1.23349	0.203095	−0.20118
9	0.79253	−0.80342	−0.52282	1.02687	0.668952	−0.21187
10	0.86120	−0.89957	−0.57538	1.13248	0.743495	−0.21589
11	1.02547	−0.88319	−0.22268	1.29774	0.390948	−0.20903
12	0.99786	−0.90313	−0.31535	1.27605	0.48725	−0.21133
13	0.87359	−1	−0.75242	1.17996	0.941667	−0.22443
14	0.93681	−0.97578	−0.57953	1.23759	0.765219	−0.21973
15	1.00349	−1	−0.49378	1.31604	0.686303	−0.2181
16	0.95417	−0.83191	−0.26556	1.20479	0.421068	−0.20926

Table 7 Weighted principal component for L_{16} experimental observations

Sl. No.	Normalized principal components X_{i1}			WPC	S/N ratio of WPC
	PCS_{11}	PCS_{21}	PCS_{31}		
1	0.999	0.278	1.000	0.728	−2.756
2	0.327	1	0.170	0.581	−4.719
3	0.078	0.234	0.321	0.137	−17.27
4	0.117	-2.9×10^{-7}	0.454	0.073	−22.74
5	0.145	0.661	0.241	0.339	−9.386
6	0.067	0.293	0.308	0.152	−16.36
7	0.015	0.355	0.227	0.143	−16.87
8	0.059	0.074	0.389	0.065	−23.77
9	0.208	0.525	0.210	0.328	−9.693
10	0.132	0.597	0.143	0.307	−10.24
11	0.013	0.256	0.258	0.105	−19.61
12	0.029	0.349	0.219	0.149	−16.51
13	0.098	0.789	2.3×10^{-5}	0.359	−8.908
14	0.056	0.618	0.079	0.268	−11.43
15	2.5×10^{-7}	0.542	0.106	0.204	−13.79
16	0.080	0.285	0.254	0.157	−16.07

Table 8 ANOVA for SN ratios

Source	DF	Seq SS	Adj SS	Adj MS	*F*	*P*
HAp volume (%)	3	52.62	52.62	17.54	2.6	0.164
Feed rate (mm/min)	3	368.84	368.84	122.946	18.25	0.004
Speed (rpm)	3	44.42	44.42	14.808	2.2	0.207
Drill diameter (mm)	1	36.91	36.91	36.907	5.48	0.066
Residual error	5	33.68	33.68	6.736	–	–
Total	15	536.47	–	–	–	–

Table 9 Response table for SN ratios

Level	HAp (vol. %)	Feed rate (mm/min)	Speed (rpm)	Drill diameter (mm)
1	−11.871	−7.686	−13.699	−12.239
2	−16.599	−10.69	−11.101	−15.276
3	−14.012	−16.886	−15.54	–
4	−12.549	−19.769	−14.69	–
Delta	4.727	12.084	4.439	3.038
Rank	2	1	3	4

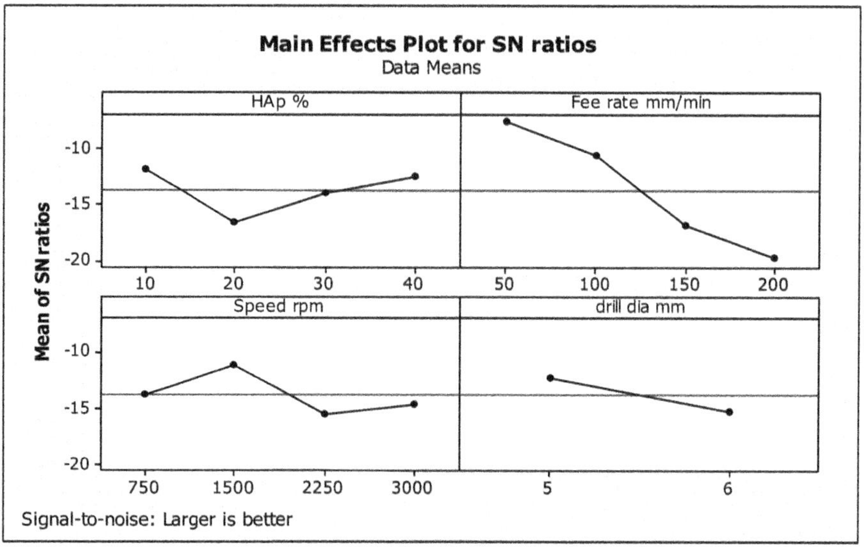

Fig. 10 Main effects plot for means of SN ratios

4.2 Estimation of Optimum Performance of Parameters

The optimum drilling parameters were analyzed by taking SN ratios of WPC, parameters are A_4, B_1, C_2, and D_1. Therefore, the predicted mean of parameters has been calculated [46] by Eq. (9)

$$S_{s/n} = \bar{Y} + (\bar{A}_1 - \bar{Y}) + (\bar{B}_1 - \bar{Y}) + (\bar{C}_2 - \bar{Y}) + (\bar{D}_1 - \bar{Y}) \tag{9}$$

where $\bar{Y}$ is the total average of *s*/*n* ratios (of MPI from Table 5), $\bar{A}_1$, $\bar{B}_1$, $\bar{C}_2$, and $\bar{D}_1$ are the *s*/*n* ratios of the average drilling parameters at the optimal levels, and $S_{s/n}$ denotes the predicted signal to noise ratio. Therefore, the calculated values of the responses are $\bar{Y} = -13.7576$, $\bar{A}_1 = -11.8713$, $\bar{B}_1 = -7.6857$, $\bar{C}_2 = -10.6873$, and $\bar{D}_1 = -12.2385$ (from Table 5). Substituting the values in Eq. (7), the predicted mean parameter is: $S_{s/n} = -1.2099$.

The confidence interval for the predicted [47] mean of SN ratio is given by Eqs. (10) and (11).

$$\text{CI} = \sqrt{F_\alpha(1; f_e).V_e\left[\frac{1}{N_{\text{eff}}} + \frac{1}{R}\right]} \tag{10}$$

$$N_{\text{eff}} = \frac{N}{1 + T_{\text{DOF}}} \tag{11}$$

where, $F_\alpha(1;f_e)$ is the F ratio required for 100 (1 − α) percent confidence interval, f_e is error degree of freedom, V_e error variance is 6.736 (from Table 7), R is the number of replication, N the number of experiment, and T_{DOF} the total degrees of freedom (from Table 6).

Substituting the values in Eq. (10), the calculated confidence interval is: CI = ±8.66809. The 95% confidence interval of the predicted optimal parameters is obtained as:

$$7.4582 < S_{s/n} < -9.8780$$

4.3 Confirmation Experiment

The confirmation experiment was performed on the parameters 10% HAp, 50 mm/min feed rate, 1500 rpm speed, and 6 mm drill diameter (i.e., $A_1B_1C_2D_2$), thus confirmation s/n ratio of WPC is −4.99678, thus which is under predicted optimal parameters.

5 Conclusions

This chapter presents a methodology using principal component analysis embedded with Taguchi method for simultaneous optimization of multiple responses of correlated in nature. The application of includes drilling of bio-composites (HAp and HDPE) extensively used in bone grafting. The optimum combination of drilling parameters can lead to minimum taper with maximum circularity at entry and exit. The optimal parameters are listed as feed rate of 50 mm/min, speed of 1500 rpm, and drill diameter of 5 mm for drilling of composites containing HAp of 10% (vol. %). Optimization of performance measures suggest PCA is a robust technique. The confirmation test shows the adequacy of the proposed methodology.

References

1. Venkatesan, J., & Kim, S. K. (2010). Effect of temperature on isolation and characterization of hydroxyapatite from tuna (Thunnusobesus) bone. *Materials, 3*, 4761–4772.
2. Fratzl, P., Gupta, H., Paschalis, E., & Roschger, P. (2004). Structure and mechanical quality of the collagen–mineral nano-composite in bone. *Journal of Materials Chemistry, 14*, 2115–2123.
3. Chowdhury, Kulkarni, A. C., Basak, A., & Roy, S. K. (2007). Wear characteristic and biocompatibility of some hydroxyapatite-collagen composite acetabular cups. *Wear, 262*(11-12), 1387–1398.
4. Tang, P., Li, G., Wang, J., Zheng, Q., & Wang, Y. (2009). Development, characterization, and validation of porous carbonated hydroxyapatite bone cement. *Journal of Biomedical Materials Research Part B, 90*, 886–893.
5. Kothamasu, R., & Haung, S. H. (2007). Adaptive Mamdani fuzzy model for condition-based maintenance. *Fuzzy Sets and Systems, 158*, 2715–2733.
6. Kmita, R. A., Slosarczyk, A., & Paszkiewicz, Z. (2006). Mechanical properties of HAp–ZrO_2 composites. *Journal of the European Ceramic Society, 26*(8), 1481–1488.
7. Park, M. S., Eanes, E. D., Antonucci, J. M., & Skrtic, D. (1998). Mechanical properties of bioactive amorphous calcium phosphate/methacrylate composites. *Dental Materials, 14*(2), 137–141.
8. Veljovic, D., Zalite, I., Palcevskis, E., Smiciklas, I., Petrovic, R., & Janakovic, D. (2010). Microwave sintering of fine-grained HAP and HAP/TCP bioceramics. *Ceramics International, 36*(2), 595–603.
9. Jaggi, H. S., Kumar, Y., Satapathy, B. K., Ray, A. R., & Patnaik, A. (2012). Analytical interpretations of structural and mechanical response of high density polyethylene/hydroxyapatite bio-composites. *Materials & Design, 36*, 757–766.
10. Ge, S., Wang, S., & Haung, X. (2009). Increasing the wear resistance of UHMWPE acetabular cups by adding natural biocompatible particles. *Wear, 267*(5–8), 770–776.
11. Deie, M., Ochi, M., Adachi, N., Nishimori, M., & Yokota, K. (2008). Artificial bone grafting [calcium hydroxyapatite ceramic with an interconnected porous structure (IP-CHA)] and core decompression for spontaneous osteonecrosis of the femoral condyle in the knee. *Knee Surgery, Sports Traumatology, Arthroscopy, 16*, 753–758.
12. Noriyukitamai, A., Ikuokudawara, T., & Hidekiyoshikawa, N. (2010). Fully interconnected porous hydroxyapatite ceramic in surgical treatment of benign bone tumor. *Journal of Orthopaedic Science, 15*, 560–568.

13. Yoshikawa, H., & Myoui, A. (2005). Bone tissue engineering with porous hydroxyapatite ceramics. *The Japanese Society for Artificial Organs, 8*, 131–136.
14. Kuriyama, K., Hashimoto, J., Murase, T., Fujii, M., Nampei, A., Hirao, M., Tsuboi, H., Myoui, A., & Yoshikawa, H. (2009). Treatment of juxta-articular intraosseous cystic lesions in rheumatoid arthritis patients with interconnected porous calcium hydroxyapatite ceramic. *Official Journal of Japan College of Rheumatology, 19*, 180–186.
15. Yoshida, Y., Osaka, S., & Tokuhashi, Y. (2009). Clinical experience of novel interconnected porous hydroxyapatite ceramics for the revision of tumor prosthesis: a case report. *World Journal of Surgical Oncology, 7*, 76.
16. Choubey, A., Chaturvedi, V., & Vimal, J. (2012). The implementation of Taguchi methodology for optimization of end milling process parameter of mild steel. *International Journal of Engineering, Science and Technology, 4*(7), 3261–3267.
17. Ross, P. J. (1995). *Taguchi techniques for quality engineering*. McGraw Hill.
18. Pahadke, M. S. (1989). *Quality engineering using robust design*. Prentice-Hall.
19. Tsao, C. C., & Hocheng, H. (2007). Parametric study on thrust force of core drill. *Journal of Materials Processing Technology, 192–193*, 37–40.
20. Kilickap, E. (2010). Modeling and optimization of burr height in drilling of Al-7075 using Taguchi method and response surface methodology. *The International Journal of Advanced Manufacturing Technology, 49*, 911–923.
21. Huang, M. S., & Lin, T. Y. (2008). Simulation of a regression-model and PCA based searching method developed for setting the robust injection molding parameters of multi-quality characteristics. *International Journal of Heat and Mass Transfer, 51*, 5828–5837.
22. Tong, L. I., Wang, C. H., & Chen, H. C. (2005). Optimization of multiple responses using principal component analysis and technique for order preference by similarity to ideal solution. *The International Journal of Advanced Manufacturing Technology, 27*, 407–414.
23. Liu, Y., Zhang, L., & Guan, Y. (2010). Sketch-based streaming PCA algorithm for network-wide traffic anomaly detection. *International Conference on Distributed Computing Systems, 1*, 807–808.
24. Salmasnia, A., Kazemzadeh, R. B., Esfahani, M. S., & Hejazi, T. H. (2013). Multiple response surface optimization with correlated data. *The International Journal of Advanced Manufacturing Technology, 64*, 841–855.
25. Sun, Y., Zeng, W., Zhao, Y., Shao, Y., & Zhou, Y. (2012). Modeling the correlation of composition-processing-property for TC11 titanium alloy based on principal component analysis and artificial neural network. *Journal of Materials Engineering and Performance, 21*, 2231–2237.
26. Bin, C. Y., & Shui, L. C. (2008). Blended coal's property prediction model based on PCA and SVM. *Journal of Central South University of Technology, 15*(2), 331–335.
27. Qiang, Z., Jun, M., & Fei, D. (2008). Optimization design of drilling string by screw coal miner based on ant colony algorithm. *Journal of Coal Science and Engineering (China), 14*(4), 686–688.
28. Barbosa, M. C., Messmer, N. R., Brazil, M., & Lobo, A. O. (2013). The effect of ultrasonic irradiation on the crystallinity of nano-hydroxyapatite produced via the wet chemical method. *Materials Science and Engineering C, 1*, 1–6.
29. Yuana, Q., Yang, Y., Chen, J., Ramuni, V., Misra, R. D. K., & Bertrand, K. J. (2010). The effect of crystallization pressure on macromolecular structure, phase evolution, and fracture resistance of nano-calcium carbonate-reinforced high-density polyethylene. *Materials Science and Engineering A, 527*, 6699–6713.
30. Wang, P., Li, C., Gong, H., Jiang, X., Wang, H., & Li, K. (2010). Effects of synthesis conditions on the morphology of hydroxyapatite nanoparticles produced by wet chemical process. *Powder Technology, 203*, 315–321.
31. Bing, A., Hi, C. X., Shun, W. F., & Ping, W. Y. (2010). Preparation of micro-sized and uniform spherical Ag powders by novel wet-chemical method. *Transactions of Nonferrous Metals Society of China, 20*, 1550–1554.

32. Saeri, M. R., Afshar, A., Ghorbani, M., Ehsani, N., & Sorrell, C. C. (2003). The wet precipitation process of hydroxyapatite. *Materials Letters, 57*, 4064–4069.
33. Velayudhan, S., Ramesh, P., Sunny, M. C., & Varma, H. K. (2000). Extrusion of hydroxyapatite to clinically significant shapes. *Materials Letters, 46*(2), 142–146.
34. Abu, B. M. S., Cheang, P., & Khor, K. A. (2003). Mechanical properties of injection molded HAp-PEEK bio-composites. *Composites Science and Technology, 63*(3-4), 421–425.
35. Wang, M., Joseph, R., & Bonfield, W. (1998). Hydroxyapatite-polyethylene composites for bone substitution: Effects of ceramic particle size and morphology. *Biomaterials, 19*(24), 2357–2366.
36. Wang, M., & Bonfield, W. (2001). Chemically coupled hydroxyapatite-polyethylene composites: Structure and properties. *Biomaterials, 22*, 1311–1320.
37. Ryan, K. R., Sproul, M. M., & Turner, C. H. (2003). Hydroxyapatite whiskers provide improved mechanical properties in reinforced polymer composites. *Journal of Biomedical Materials Research Part A, 67*, 801–812.
38. Fang, L., Leng, Y., & Gao, P. (2006). Processing and mechanical properties of HA/UHMWPE Nano composites. *Biomaterials, 27*, 3701–3707.
39. Liu, Y., Zhang, L., & Guan, Y. (2010). Sketch-based streaming PCA algorithm for network-wide traffic anomaly Detection. *International Conference on Distributed Computing System, 2010*, 807–816.
40. Ali Salmasnia, A., & Kazemzadeh, R. B. (2013). Multiple response surface optimization with correlated data. *The International Journal of Advanced Manufacturing Technology, 64*, 841–855.
41. Chakravorty, R., Gauri, S. K., & Chakraborty, S. (2012). Optimization of correlated responses of EDM process. *Materials and Manufacturing Processes, 27*, 337–347.
42. Kilickap, E. (2010). Determination of optimum parameters on delamination in drilling of GFRP composites by Taguchi method. *Indian Journal of Engineering & Materials Sciences, 17*(4), 265–274.
43. Biswas, R., Kuar, A. S., Sarkar, S., & Mitra, S. (2010). A parametric study of pulsed Nd:YAG laser micro-drilling of gamma-titanium aluminide. *Journal of Optics and Laser Technology, 42*(1), 23–31.
44. Ajaal, T. T., & Smith, R. W. (2008). Employing the Taguchi method in optimizing the scaffold production process for artificial bone graft. *Journal of Materials Processing Technology, 209*(4), 1521–1532.
45. Gopalsamy, B. M., Mondal, B., & Ghosh, S. (2009). Taguchi method and ANOVA: An approach for process parameters optimization of hard machining while machining hardened steel. *Journal of Scientific and Industrial Research, 68*(8), 686–695.
46. Chaulia, P. K., & Das, R. (2008). Process parameter optimization for fly ash brick by Taguchi method. *Materials Research, 11*(2), 159–164.
47. Singh, H., & Kumar, P. (2004). Tool wear optimization in turning operation by Taguchi method. *Indian Journal of Engineering & Materials Sciences, 11*(1), 19–24.

A Framework for Magnetic Field-Assisted Electrical Discharge Machining (MFA-EDM) of Inconel-625 Using Bio-oil Dielectric

Mohd Yunus Khan, P. Sudhakar Rao, and B. S. Pabla

Nomenclature

3-DP	3-D printing
ECM	Electrochemical machining
EDS	Energy dispersive X-ray spectroscopy
EDM	Electrical discharge machining
EWR	Electrode wear rate
HMP	Hybrid machining process
MFA-EDM	Magnetic field-assisted EDM
MFAPM-EDM	Magnetic field-assisted powder-mixed-EDM
MMC	Metal matrix composite
MRR	Material removal rate
RLT	Recast layer thickness
SEM	Scanning electron microscopy
SR	Surface roughness
TWR	Tool wear rate

1 Introduction

In traditional machining process, material expulsion from the workpiece occurs as chips; hence, it is hard to accomplish required accuracy. Such accuracy can be easily achieved by employing non-traditional methods [1]. These machining processes are used where traditional ones are unattainable, non-acceptable, or costly. It is

M. Y. Khan (✉) · P. S. Rao · B. S. Pabla
Department of Mechanical Engineering, National Institute of Technical Teachers Training & Research, Chandigarh, India
e-mail: yunus.mech19@nitttrchd.ac.in

C. Prakash et al. (eds.), *Additive, Subtractive, and Hybrid Technologies*, Mechanical Engineering Series, https://doi.org/10.1007/978-3-030-99569-0_6

extremely difficult to work on hard and brittle materials conventionally. If such materials are worked using traditional process, either tool wears or the work sample failures. Various nontraditional methods have been developed to accomplish exceptional machining situations. These processes find wide applicability in various areas of manufacturing. However, non-traditional processes are costly and require of high-skilled operator and complex equipment [2–4]. Some of the nontraditional methods are electrical discharge machining, electro-chemical machining, abrasive jet machining, laser beam machining, electron beam machining, etc. Effectual utilization of the capacity of a non-traditional process needs careful selection of the suitable method for a particular application [5]. EDM is used for hard [6] and difficult to machine materials [7]. Spark machining is well received since more than 70 years [2, 3]. It has achieved acknowledgement around the world and supplanted traditional methods of machining [8]. Chief advantage of EDM process is that tool and work sample are not contacted physically. EDM process is thus free from complexities such as mechanical stresses [9]. Spark machining is usually preferred suitable for a manufacture of complex, precise and smaller products [10]. Variations of EDM includes die-sinking, wire-cut, etc.

2 Hybrid Machining Process (HMP)

HMP is a combination of various machining methods for machining work piece with a higher machining efficiency. Hybrid techniques use the combined benefits of at least two individual methods. HMP additionally diminishes adverse effects of individual process [2, 3]. With the headway in the field of HMP, productivity has greatly enhanced. By employing HMP, it is conceivable to accomplish machining of the newer materials, and development of complex shape is possible with improved precision. There are three types of HMP.

1. Assisted hybrid method in which the material removal happens by one source and the other source just helps it. For example, MFA-EDM.
2. Combined hybrid method in which different material evacuation techniques at the same time work on the work sample and participate in material evacuation. For example, electro-chemical discharge machining.
3. Controlled application of method in which various techniques are combined on the same machine. Such system lessens the process chain. For example, grind hardening [11, 12].

HMP shortens the existing techniques by combining at least two processes on the same setup accordingly, eliminating issues of referencing, clamping, alignment, etc. HMP incorporates the advantages and process capabilities of various methods, and better machining performance is achieved. Traditional methods are restricted by the tool strength and chip formation. Thus, these methods fail to machine advanced materials. Even non-traditional methods have limitation of lower material removal rate. Therefore, to fulfill challenging machining needs, two or more process are combined together [11, 12]. One of the major limitations of HMP is high cost which includes initial and operational cost of the setup. These methods are in their outset yet have the capability of turning into predominant power in the manufacturing sector. Combining advantages of two or more processes at one setup can be quite beneficial in improving productivity. System can be operated as individual process or hybrid process as and when required. The way to advancement for HMP will be optimizing the manufacturing technique to impart flexibility.

3 Framework for Future Research

Several investigations have been carried out on tool electrode [13], flushing [14], dielectric medium [15], mixing of powder [16], and so on. The goal of this review work was to understand the concept of EDM hybrid process and its effects on EDM performance. Review of investigations done on magnetic field-assisted EDM process has been conducted and presented in systematic manner so as to give research progress in this field which can be effectively used for future research. A framework for MFA-EDM process of Inconel-625 with bio-oil dielectric is presented. Experiments will be conducted to study surface modification through material deposition using variable density electrodes developed by 3-D printing technology. Important parameters to be considered in future investigation are summarized in Table 1. All experimentations will be conducted on sinker EDM (Make: Toolcraft India Limited; Model G30) installed at National Institute of Technical Teachers' Training and Research, Chandigarh. The pictorial view of EDM is shown in Fig. 1.

Table 1 Various parameters selected for experimentation

Input variables	Peak current, pulse on-time and pulse off-time
Constant variables	Gap, voltage, flushing pressure
Electrodes	Cu electrode of three different densities
Di-electric fluids	EDM oil and bio-oil
Magnetic field strengths	Three different strengths
Output parameters (machining)	Material removal rate, tool wear rate, surface roughness, and micro-hardness
Characteristic study	Scanning electron microscopy (SEM) and energy dispersive X-ray spectroscopy (EDS)
Output parameters (surface modification)	Material deposition rate, tool wear rate, surface roughness, micro-hardness, and layer thickness

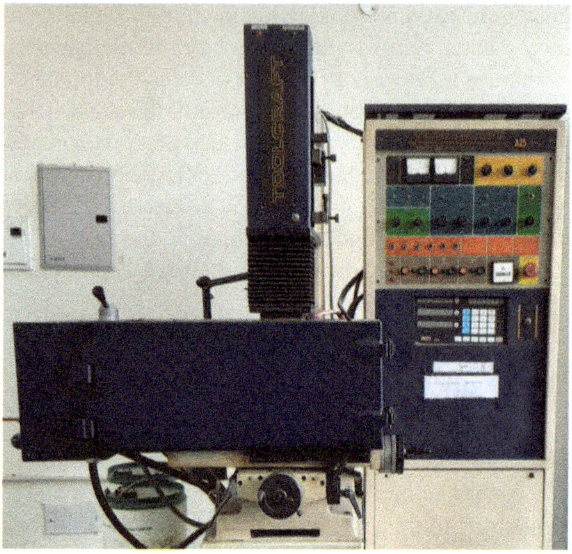

Fig. 1 Photograph of EDM (Courtesy: National Institute of Technical Teachers' Training and Research, Chandigarh)

4 Studies on Hybrid EDM Process

Non-traditional machining methods have its advantages and limits. To outperform associated constraints, additional method is added to hybridize it. This not only reduces the short comings but also maximizes the benefits. In this manner, it very well may be said that the hybridized process has advantages of the each of the process. HMP finds applicability in producing different tools, machine parts, electronic components, and micro-machine components because of high accuracy and surface quality. Hybrid EDM process can be employed for proto-typing or single-part production. Industrial sectors receiving prompt benefit of HMP include aircraft, health, and mold and die making.

Zhang et al. [17] researched on ultrasonic-assisted EDM and found increase in MRR. Yadav et al. [18] assessed electro-discharge diamond grinding process and noticed that MRR increased with increase in current, wheel speed and on time and decreased with high surface finish mode of EDM. Nguyen et al. [19] analyzed combined micro-EDM and micro-ECM in order to achieve improvement in surface finish and machining efficiency. Yadav and Yadava [20] examined on slotted electrical discharge diamond grinding and noted that higher wheel RPM results in high MRR and superior surface finish. Li et al. [21] assessed combined EDM and end milling and observed an improvement in the machining efficiency. An improvement in SR and TWR was noted. Al-Ahmari et al. [22] analyzed the combination of laser and micro-EDM drilling process and found that as a result of combined process, machining time reduced and MRR increased without impacting the quality of micro-hole. Lin et al. [23] evaluated the characteristics of gas medium EDM combined with abrasive jet machining and ultrasonic vibration. Higher MRR was obtained with combined process. The ultrasonic vibration-integrated HMP decreased recast layer thickness (RLT). In another study, Lin et al. [24] studied the characteristics of gas medium EDM coupled with abrasive jet machining. Higher machining efficiency was observed in the combined process. Ni et al. [25] developed a hybrid ultrasonic EDM setup that can use these techniques concurrently. Results showed more stable machining and higher MRR. Zhang and Xie [26] investigated on ultrasonic-assisted EDM of micro-hole arrays with tool electrode arrays. They suggested that decrease in concentration of debris particles is an efficacious method for achieving improvement in the performance of EDM.

5 Mechanism of Magnetic Field-Assisted EDM (MFA-EDM) Process

In MFA-EDM technique, a pair of rectangular permanent magnets is employed to create magnetic field around the machining zone. These magnets are placed around the tool–work piece intersection with certain distance between their faces. Numerous magnets can be employed to develop required magnetic field intensity. The magnetic fields can possibly impact plasma behavior and had been a focal point of interest to plasma research since the 1950s [27]. Investigation conducted by Joshi et al. [28] showed that a resultant force is produced by the vector summation of electric force and magnetic force, which acts on electrons. The electrons, in this manner, are directed toward the spark zone affected by the resultant force, which consequently encourages the start of Lorentz forces in the discharge zone. These forces are responsible for increase in the electron density and constraining electron motion. Lorentz forces influence plasma and consequently MRR. MFA-EDM is relatively a new idea for machining of non-magnetic workpiece. Utilization of a magnetic field also improves the plasma density and results in an increase in the stability of plasma [29].

6 Studies on MFA-EDM Process

Researchers worldwide have studied feasibility of applying magnetic field in achieving refinement of EDM operation. Spark column in EDM comprises of charged particles which includes electrons and ions, movable in the line of electrical field in a linear track. When magnetic field is employed, particles change their track, and EDM performance is altered.

Kim et al. [30] studied the influence of magnetic field on an electrolytic finishing operation. They obtained an enhanced efficiency with high-quality surface finish while working on SM45C steel work piece. Yamaguchi and Shinmura [31] examined the applicability of magnetic field-assisted finishing of alumina ceramic work piece. They observed smoother the inner surface and to improve the form accuracy of ceramic tube. Pa [32] investigated magnetically assisted finishing technique for efficient electrode gap dreg removal during electro-chemical finishing of cylindrical surfaces. Results indicated smooth surface of SKD1 alloy tool steel workpiece. Chattopadhyay et al. [33] induced magnetic field during rotary EDM during EN-8 steel work piece. Reduction in machining cost was achieved. MRR and EWR increased and decreased, respectively. Lin and Lee [34] examined MFA-EDM of SKD 61 steel and found that magnetic force resulted in rapid and easy material evacuation in gap and intensified the process on uniqueness by providing stability. Joshi et al. [28] investigated the effect of pulsating magnetic field during machining stainless steel for achieving improvement in EDM process. It resulted in higher plasma ionization and steep confinement, thus supporting in higher heat transfer and better MRR. Khan et al. [35] studied the impact of external magnetic field on EDM of SUS 304 stainless steel. They found that introduction of magnetic field enhances material evacuation away from the machining gap and improves SR.

Reddy et al. [36] used permanent magnets at several junctions around the spark zone for achieving improvement in surface quality of the high carbon high chromium steel workpiece. Better surface finish was seen when process variables were set at optimum value with magnetic field. Rouniyar and Shandilya [37] experimented on Al6061 alloy with magnetic field-assisted powder-mixed-EDM (MFAPM-EDM) operation. They obtained higher MRR and lower TWR with powder mixing and magnetic field. Kumar et al. [38] performed EDM operation of Inconel 706 using the combination of powder mixing and magnetic field. Surface morphology showed the formation of melted debris, micro-holes, and micro-globules on its surface. Ming et al. [39] investigated MAF-EDM on magnetic and non-magnetic materials. MRR for magnetic materials improved more drastically as compared to non-magnetic materials. Rouniyar and Shandilya [40] experimented on using a combination of Al powder and magnetic field on Al6061 alloy on EDM. Improvements in SR and RLT were noted with MFAPM-EDM process. Surface morphology showed less cracks, voids, crater and molten debris particles on the machined surface.

7 Inferences from Literature

Following inferences can be drawn from this study are:

1. Overview of hybrid machining process along with its mechanism, classification, and advantages is presented. Comprehensive review of researches on hybrid EDM process has been presented. Experimentation has been conducted on complicated hybrid machining processes. Hybrid EDM process is achieved by providing an attachment on the conventional EDM.

2. Combination of EDM and other process resulted in improvement in MRR and SR. Better surface finish and machining efficiency were achieved. Although, the setup becomes complicated.
3. HMP can reduce the process chains, improve productivity, and increase machining efficiency by coupling two (or more) techniques. Such a process can have advantages of both the processes.
4. Challenges in the implementation of hybrid processes have also been discussed. Main challenges include designing of the machining equipment and its cost.
5. Available literature on magnetic field-assisted EDM process has been reviewed. MFA-EDM assumes an indispensable role in increasing MRR, improving surface finish, and decreasing TWR. MFA-EDM also facilitated the process stability.
6. No work on hybridized electric discharge machining with bio-oil as dielectric fluid is found during literature review. Effect of magnetization and use of high-frequency ultrasonic vibrations need to be investigated.

8 Research Gaps

The following research gaps have been observed after conducting literature review:

1. Magnetic field/ultrasonic-assisted EDM with bio-oil dielectric has not been explored by any researcher. The process of hybrid machining utilizes the advantages of the individual processes.
2. Research can be conducted on the use of vegetable oil hybrid dielectric (i.e., combination of bio-oil with other dielectric fluid) during hybrid EDM process.
3. No researchers have used electrode developed by 3D printing technique in MFA-EDM process. 3-DP is a computer-controlled additive manufacturing technique of developing products by laying down material layer by layer. Through this process, it is possible to develop variable density electrodes. Influence of variable density of electrode on surface deposition can be taken up.

9 Steps to Be Followed

The following steps will be followed for further research work and are represented by flowchart in Fig. 2.

1. The major target of the current research work will be to study the impact of process parameters on different output parameters during MFA-EDM process of Inconel 625 using copper electrode and bio-oil dielectric during machining.
2. Inconel alloy will be selected for its wide applicability and for its specific properties. Inconel-625 is known for its resistance to corrosion and oxidation and its capacity to keep up their structural integrity at elevated temperature.
3. Different density electrodes of copper will be developed by 3-DP technique.

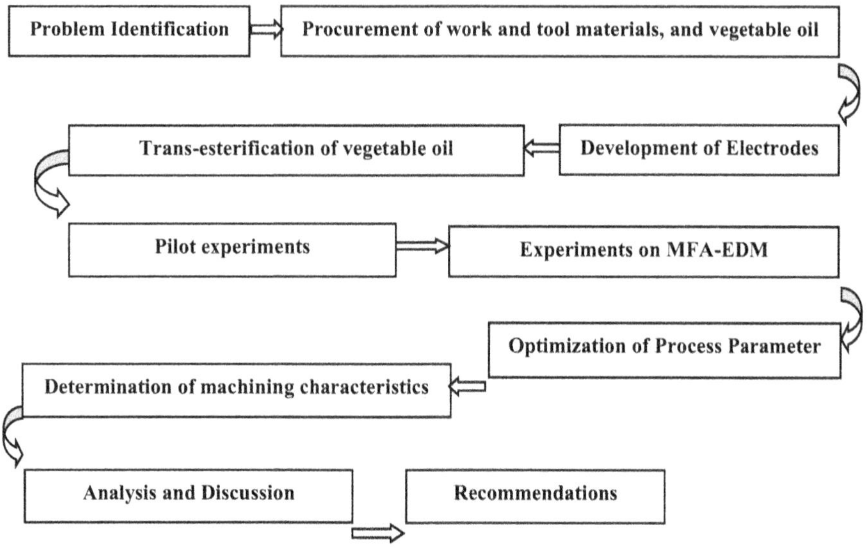

Fig. 2 Flowchart representing steps to be followed in this work

4. An attachment of permanent magnets will be developed for creating required magnetic field intensity.
5. Influence of process parameters will be taken into account, and several levels of these parameters will be considered for doing experiments on die-sinking EDM.
6. Transesterified vegetable oil dielectric will be used for conducting experiments.
7. Among the process parameters, peak current, pulse on-time, and pulse off-time will be considered while material removal rate, tool wear rate, and surface roughness will be selected as output parameters.
8. Study of surface modification through deposition of material will be conducted using variable density electrode with negative polarity condition. Effect of electrode density on surface properties of machined components will be investigated.
9. The process parameters will be optimized for the best performance of EDM.

10 Conclusions

Electric discharge machining process finds wide applications in various industries. In order to improve performance of conventional EDM, it is hybridized with other machining process(es). In this chapter, a framework for conducting experiments using magnetic field-assisted EDM of Inconel-625 is presented. Conclusions drawn from this study are given below.

- Summary of studies on hybrid EDM and MFA-EDM processes is presented.

- MFA-EDM process plays a significant role in improving MRR and surface finish while decreasing TWR. The process of MFA-EDM has the ability of enhancing the spark through Lorentz force.
- Experiments on die sinker EDM are proposed using bio-oil dielectric. Thus, a combined study of using magnetic field-assisted EDM with bio-oil dielectric is suggested.
- EDM technology is also used for material deposition. Further research work is proposed on surface modification using variable densities of Cu electrodes synthesized using 3-D printing technique.

References

1. Rizwee, M., Rao, P. S., & Khan, M. Y. (2020). Recent advancement in electric discharge machining of metal matrix composite materials. *Materials Today: Proceeding, 37*, 2829–2836.
2. Khan, M. Y., & Rao, P. S. (2019). Electrical discharge machining: Vital to manufacturing industries. *International Journal of Innovative Technology and Exploring Engineering, 8*(11), 1696–1701.
3. Khan, M. Y., & Rao, P. S. (2019). Hybridization of electrical discharge machining process. *International Journal of Engineering and Advanced Technology, 9*(1), 1059–1065.
4. Pandey, P. C., & Shan, H. S. (1999). *Modern machining process*. Tata McGraw-Mill Publishing Company Ltd..
5. Khan, M. Y., Rao, P. S., & Pabla, B. S. (2020). *Review of electrical discharge machining process with nanopowder and CNT mixed dielectric fluid*. In Proceedings of 4th international online multidisciplinary research conference (IOMRC-2020), Hyderabad, October 23.
6. Gadalla, A. M., & Bedi, H. S. (1991). Machining of TiB_2 and its composites. *Materials and Manufacturing Processes, 6*(1), 125–137.
7. Gua, C., Di, S., & Wei, D. (2016). Study of electrical discharge machining performance in water-based working fluid. *Materials and Manufacturing Processes, 31*(14), 1865–1871.
8. Thakur, A., Rao, P. S., & Khan, M. Y. (2020). Study and optimization of surface roughness parameter during electrical discharge machining of titanium alloy (Ti-6246). *Materials Today: Proceedings, 44*, 11.
9. Sen, I., Karthikeyan, G., Ramkumar, J., & Balasubramaniam, R. (2012). A study on machinability of b-modified Ti-6Al-4V alloys by EDM. *Materials and Manufacturing Processes, 27*(3), 348–354.
10. Shin, H. S., Park, M. S., Kim, B. H., & Chu, C. N. (2011). Recent researches in micro electrical machining. *International Journal of Precision Engineering and Manufacturing, 12*(2), 371–380.
11. Singh, B. P., Sidhu, S. S., & Payal, H. S. (2018). Investigation of magnetic field-assisted EDM of composites. *Materials and Manufacturing Processes, 33*(6), 670–675.
12. Singh, K., Maurya, S. K., & Kumar, S. (2018). A review of introduction to hybrid machining process. *International Journal of Advance Engineering and Technology, 2*(3), 22–26.
13. Singh, J., Singh, G., & Pandey, P. M. (2021). Electric discharge machining using rapid manufactured complex shape copper electrode with cryogenic cooling channel. *Proceedings of the Institution of Mechanical Engineers, Part B: Journal of Engineering Manufacture, 235*(1-2), 173–185.
14. Zhang, M., Jiang, X., Feng, Y., Jia, B., & Wang, J. (2021). Effects of workpiece polarity and flushing pressure on the arc plasma and the crater of single-pulsed arc discharges. *The International Journal of Advanced Manufacturing Technology, 112*, 1593–1600.

15. Singaravel, B., Shekar, K. C., Reddy, G. G., & Prasad, S. D. (2020). Experimental investigation of vegetable oil as dielectric fluid in electric discharge machining of Ti-6Al-4V. *Ain Shams Engineering Journal, 11*(1), 143–147.
16. Ishfaq, K., Asad, M., Anwar, S., Pruncu, C. I., Saleh, M., & Ahmad, S. (2021). A comprehensive analysis of the effect of graphene-based dielectric for sustainable electric discharge machining of Ti-6Al-4V. *Materials, 14*(1), 23.
17. Zhang, Q. H., Du, R., Zhang, J. H., & Zhang, Q. B. (2006). An investigation of ultrasonic-assisted electrical discharge machining in gas. *International Journal of Machine Tools and Manufacture, 46*(12-13), 1582–1588.
18. Yadav, S. K. S., Yadava, V., & Narayana, V. L. (2008). Experimental study and parameter design of electro-discharge diamond grinding. *The International Journal of Advanced Manufacturing Technology, 36*(1-2), 34–42.
19. Nguyen, M. D., Rahman, M., & Wong, Y. S. (2012). Simultaneous micro-EDM and micro-ECM in low-resistivity deionized water. *International Journal of Machine Tools and Manufacturing, 54*, 55–65.
20. Yadav, R. N., & Yadava, V. (2014). Machining performance of slotted-electrical discharge diamond face grinding of Al/SiC/Gr composite. *Materials and Manufacturing Processes, 29*(5), 585–592.
21. Li, C. P., Kim, M. Y., Islam, M. M., & Ko, T. J. (2016). Mechanism analysis of hybrid machining process comprising EDM and end milling. *Journal of Materials Processing Technology, 237*, 309–319.
22. Al-Ahmari, A. M. A., Rasheed, M. S., Mohammed, M. K., & Saleh, T. (2016). A hybrid machining process combining micro-EDM and laser beam machining of nickel-titanium-based shape memory alloy. *Materials and Manufacturing Processes, 31*(4), 447–455.
23. Lin, Y. C., Hung, J. C., Chow, H. M., Wang, A. C., & Chen, J. T. (2016). Machining characteristics of a hybrid process of EDM in gas combined with ultrasonic vibration and AJM. *Procedia CIRP, 42*(1), 167–172.
24. Lin, Y. C., Hung, J. C., Lee, H. M., Wang, A. C., & Fan, S. F. (2018). Machining performances of electrical discharge machining combined with abrasive jet machining. *Procedia CIRP, 68*, 162–167.
25. Ni, H., Gong, H., Dong, Y. H., Fang, F. Z., & Wang, Y. (2018). A comparative investigation on hybrid EDM for drilling small deep holes. *The International Journal of Advanced Manufacturing Technology, 95*(1-4), 1465–1472.
26. Zhang, Y., & Xie, B. (2021). Investigation on hole diameter non-uniformity of hole arrays by ultrasonic vibration-assisted EDM. *The International Journal of Advanced Manufacturing Technology, 112*, 2.
27. Sternberg, N., Godyak, V., & Hoffman, D. (2006). Magnetic field effects on gas discharge plasmas. *Physics of Plasmas, 13*(6), 063511.
28. Joshi, S., Govindan, P., Malshe, A., & Rajurkar, K. (2011). Experimental characterization of dry EDM performed in a pulsating magnetic field. *CIRP Annals, 60*(1), 239–242.
29. Kim, K. S. (2009). Influence of a transverse magnetic field on arc root movements in a dc plasma torch: Diamagnetic effect of arc column. *Applied Physics Letters, 94*(12), 121501.
30. Kim, J. D., Jin, D. X., & Choi, M. S. (1997). Study on the effect of a magnetic field on an electrolytic finishing process. *International Journal of Machine Tools and Manufacture, 37*(4), 401–408.
31. Yamaguchi, H., & Shinmura, T. (2004). Internal finishing process for alumina ceramic components by a magnetic field assisted finishing process. *Precision Engineering, 28*(2), 135–142.
32. Pa, P. S. (2009). Magnetic assistance in cylinder-surfaces finish. *Materials and Manufacturing Processes, 24*(7-8), 819–823.
33. Chattopadhyay, K. D., Satsangi, P. S., Verma, S., & Sharma, P. C. (2008). Analysis of rotary electrical discharge machining characteristics in reversal magnetic field for copper-en8 steel system. *The International Journal of Advanced Manufacturing Technology, 38*(9-10), 925–937.

34. Lin, Y. C., & Lee, H. S. (2008). Machining characteristics of magnetic force-assisted EDM. *International Journal of Machine Tools and Manufacture, 48*(11), 1179–1186.
35. Khan, A. A., Ndaliman, M. B., Ali, M. Y., & Al-Falahi, M. D. (2013). *The effect of EDM with external magnetic field on surface roughness of stainless steel*. In Proceedings of the 2nd international conference on mechanical, automotive and aerospace engineering, Kuala Lumpur, Malaysia, July 2–4.
36. Reddy, C. K., Hussain, M. M., Satyanarayana, S., & Krishna, M. M. (2018). Experimental investigation–magnetic assisted electro discharge machining. *IOP Conference Series: Materials Science and Engineering, 346*(1), 012062.
37. Rouniyar, A. K., & Shandilya, P. (2019). Fabrication and experimental investigation of magnetic field assisted powder mixed electrical discharge machining on machining of aluminum 6061 alloy. *Proceedings of the Institution of Mechanical Engineers, Part B: Journal of Engineering Manufacture, 233*(12), 2283–2291.
38. Kumar, S., Goud, M., & Suri, N. M. (2020). An investigation of magnetic-field-assisted EDM by silicon and boron based dielectric of Inconel 706. *Silicon, 2020*, 1–9.
39. Ming, W., Shen, F., Zhang, Z., Huang, H., Du, J., & Wu, J. (2020). A comparative investigation on magnetic field-assisted EDM of magnetic and non-magnetic materials. *The International Journal of Advanced Manufacturing Technology, 109*(3), 1103–1116.
40. Rouniyar, A. K., & Shandilya, P. (2020). Experimental investigation on recast layer and surface roughness on aluminum 6061 alloy during magnetic field assisted powder mixed electrical discharge machining. *Journal of Materials Engineering and Performance, 29*(12), 7981–7992.

Investigation of Mechanical Properties in Friction Stir Welded Mg AZ 31 Alloy Workpieces

N. Giri, G. S. Brar, and A. S. Shahi

1 Introduction: Background and Driving Forces

Friction stir welding (FSW) is a solid-state material processing strategy dependent on the standards of FSW developed by The Welding Institute (TWI) in 1991 [1]. This welding process uses the heat generated by friction while rubbing two pieces. Frictional heat energy raises the temperature of the welding surface; then after sufficient pressure is applied, diffusion process will start. Finally, diffusion welding of the joint takes place.

In this welding process, two flat surface pieces are joined by a tool that must be round in shape, so that it can be subjected to rotation. The round-shaped tool is rotated with the help of an external rotating machine. Pressure is exerted in the direction of the rubbing surfaces. In this condition, high heat is generated which increases the temperature of the material pieces, and diffusion starts. This type of welding is used for joining similar or dissimilar pieces of material such as aluminum to magnesium alloy.

This welding uses a non-consumable tool to join workpieces without melting. This is a solid-state type of joining measure. The heat-affected zone is common as

The original version of this chapter was revised. The correction to this chapter is available at https://doi.org/10.1007/978-3-030-99569-0_14

N. Giri · G. S. Brar (✉)
Department of Mechanical Engineering, National Institute of Technology,
Srinagar, Uttarakhand, India
e-mail: brar.gurinder@nituk.ac.in

A. S. Shahi
Department of Mechanical Engineering, Sant Longowal Institute of Engineering
and Technology, Sangrur, Punjab, India

C. Prakash et al. (eds.), *Additive, Subtractive, and Hybrid Technologies*,
Mechanical Engineering Series, https://doi.org/10.1007/978-3-030-99569-0_7

Fig. 1 Friction stir welding setup for joining 2 Mg AZ 31 alloy workpieces

demonstrated by the name and is exposed to thermal cycle during process but cannot deform. HAZ is lower due to less temperature rise but still significant. Heat is generated between wear tool due to resistance and the workpiece. This heat softens the material without melting the material. Now due to the motion of the tool forward, plastic deformation of the workpiece takes place [3, 5].

The FS welding machine has a limit of joining workpiece thickness from 0.5 mm to 10 mm. Most of the available literature on friction stir welding joint was on workpiece thickness ranging from 6 mm to 8 mm, but limited work was done on welding of workpiece of thickness below 5 mm [4]. The effort in this work was to weld and test specimen of thickness 3 mm, and pictorial view of a friction stir welding machine is shown in Fig. 1.

The welded joint is prepared by extreme plastic deformation of the material and blending activity forced by the friction stir processing tool. The FSP has also been used for basic miniature change and homogenization of cast alloys and powder metallurgy manufactured parts [1, 6]. The tool revolution and straight interpretation force lead the tool probe to shear material from the advancing side (AS) and move it to the retreating side (RS). Altogether, the tool shoulder provides forging action to confine ejecting out of the processed material. The combined thermo-mechanical stirring effect leads to be distinguished as friction stir zone (FSZ)/nugget zone, thermo-mechanically influenced area (TMAZ), and heat-affected zone (HAZ) [6].

2 Research Methods

The detailed experimental procedure used for joining two similar or dissimilar materials by friction stir welding technique is shown in Fig. 2. The magnesium AZ 31 workpieces were welded by friction stir welding process as shown in Fig. 3.

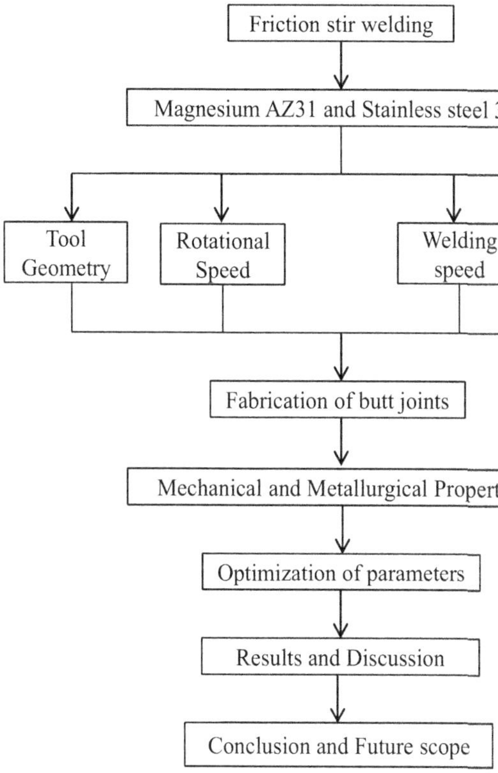

Fig. 2 The detailed experimental procedure used for joining two similar or dissimilar materials by friction stir welding technique

3 Welded Joint Characterization

The welded workpieces were tested for tensile strength, flexural strength, and impact toughness.

3.1 Tensile Strength

The samples were cut according to ASTM E8 having dimensions of 100 mm × 10 mm × 3 mm as shown in Fig. 4. The speed of the cross head was kept at 2 mm/min. The span length of the specimen was 100 mm. The sample was loaded at a pace of 1 KN/min according to ASTM designations. Figure 5 shows the tensile specimens before and after tensile strength testing with fracture location.

Fig. 3 The magnesium AZ 31 workpieces being welded by friction stir welding process

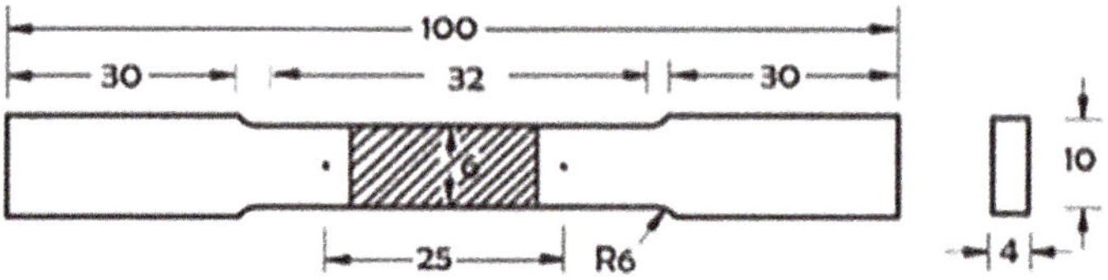

Fig. 4 The samples prepared according to ASTM E8 having dimensions of 100 mm × 10 mm × 3 mm for tensile testing

3.2 *Flexural Strength*

The flexural test was also done on Universal Testing Machine by using code ASTM B577 Samples were cut in 100 mm × 10 mm × 3 mm as shown in Fig. 6. Three-point forces were exerted, where one force was exerted from the up side and the other two forces were exerted from the down side.

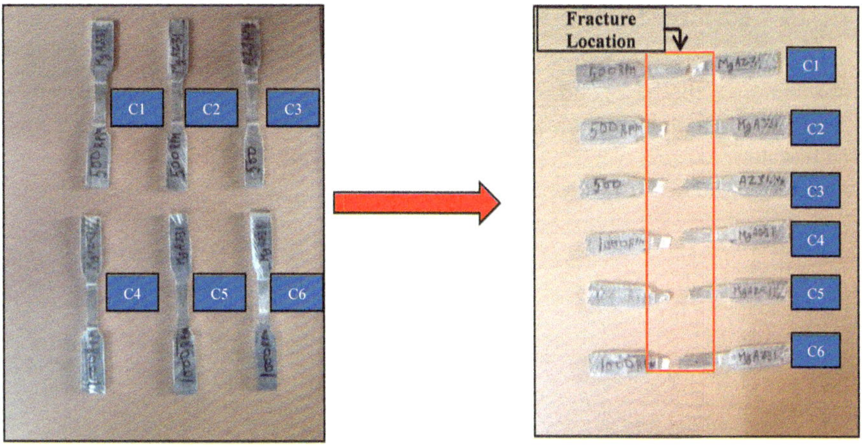

Fig. 5 The tensile specimens before and after tensile strength testing with fracture location

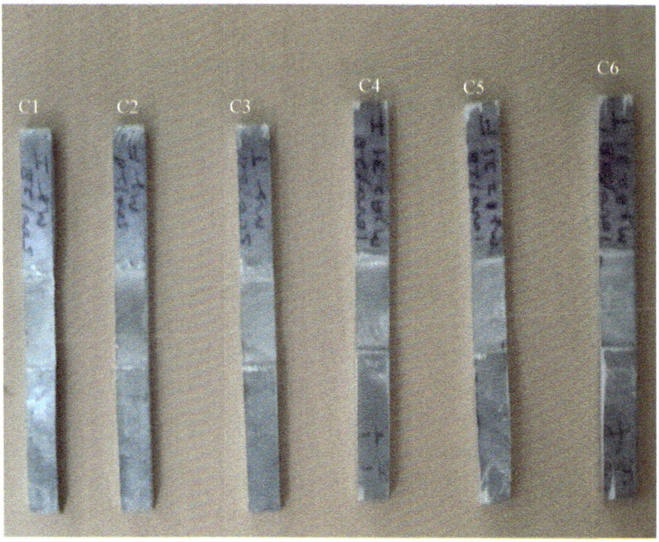

Fig. 6 The samples prepared for flexural test to be carried out on Universal Testing Machine as per ASTM B577 standard

3.3 Impact Toughness

For the present research study, ASTM E23 was utilized to measure the impact toughness and energy of the fabricated welded joints. The welding joints were cut in 55 mm × 10 mm × 3 mm.

4 Results and Discussion

The results of the best conducted on the fabricated welded joint such as a tensile strength and flexural strength have been shown. The FSW technique factors control the metallurgical and mechanical conduct of weldments. This part depicts the after-effect of various FSW strategy factors on mechanical and microstructural conduct of the joints. The effects of various cycle factors on the ideal response have been clarified graphically for better understanding. It inspects the comprehension with respect to the relationship of various friction stir welding process parameters taken for the examination and their impact on the joint properties.

4.1 Tensile Strength

Tensile strength and the percentage of elongation of the similar FSW joints has been evaluated for two tool rotation velocities. Three samples are tried for every rotation velocity, and the mean of the outcomes is shown in Table 1. All welded joints have lower tensile strength; percentage of elongation contrasted with the base metal of magnesium AZ 31. All welded joints are broken along the retreating side.

The tensile strength of the samples (C-4 to C-6) increased from 5% to 19% with increasing the tool rotation velocity to 1000 rpm as compared with the samples (C-1 to C-3) with rotational speed 500 rpm as shown in Fig. 7. It may be due to more heat generated in the welded zone, so the percentage of elongation is high in samples of rotational speed 1000 rpm as compared to the samples of rotation speed 500 rpm shown in Fig. 8. The tensile strength of the samples increased with increasing the tool rotation velocity as shown in Table 1.

The impact of the welding velocity on the UTS and percentage of elongation properties of the friction stir welding welded joints is induced through Fig. 7. Among the two tool rotational velocities, 1000 rpm shows prevalent tractable properties. These shows a rise in tool rotational velocity raise the tensile properties of a welded joint. The fundamental idea of utilization of higher rotational velocity is to develop high heat input, as a result, welded joint shows good tensile strength.

Table 1 Ultimate tensile strength and percentage of elongation in FSW workpieces due to varying rotational velocity

Tool rotation velocity (rpm)	Welding velocity (mm/min)	UTS (MPa)			Avg. UTS (MPa)	% of elongation			Avg. % of elongation
		Trial 1	Trial 2	Trial 3		Trial 1	Trial 2	Trial 3	
500	28	181.3	174.6	165.1	173.4	5.38	7.12	8.45	6.98
1000	28	190.2	202.3	217.6	203.3	9.26	10.38	13.08	10.9

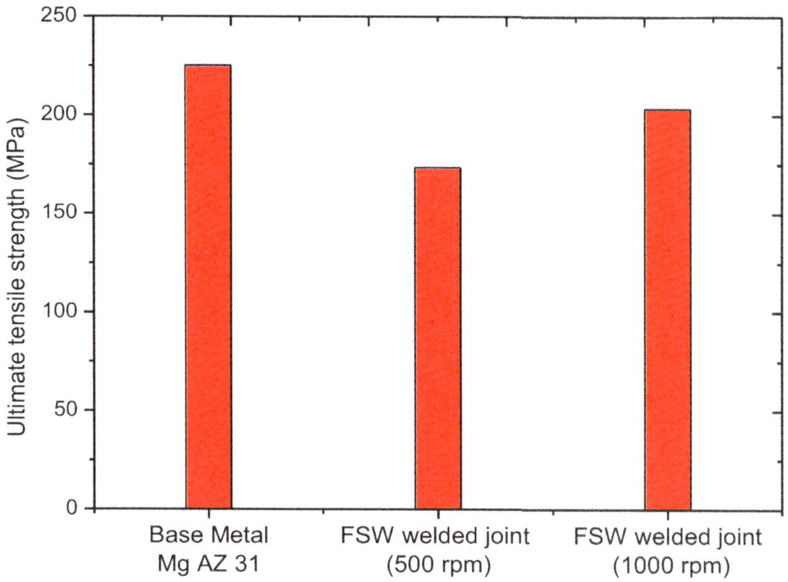

Fig. 7 The impact of the welding rotational speed on the ultimate tensile strength in the friction stir welded workpieces

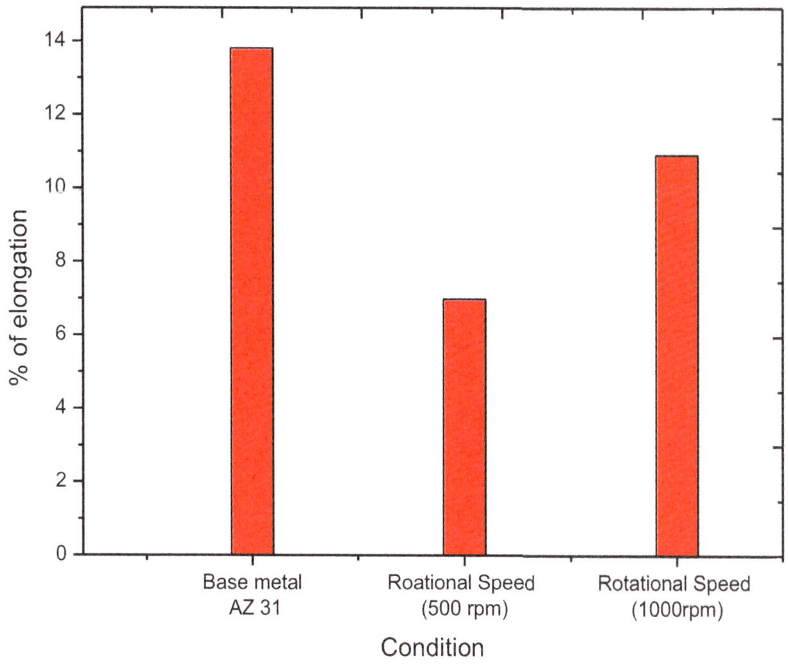

Fig. 8 The impact of the welding rotational speed on the percentage of elongation in the friction stir welded workpieces

Table 2 Compressive strength in FSW workpieces due to varying rotational velocity

Tool rotation velocity (rpm)	Welding velocity (mm/min)	Compressive strength (MPa)			Avg. compressive strength (MPa)
		Trial 1	Trial 2	Trial 3	
500	28	11.1	13.3	15.1	13.1
1000	28	10.3	11.8	10.9	11

These uncover that the ductility of stir area is lower than that of the base metals. Many researchers have found similar trends on increase in the tensile strength and percentage of elongation with increasing the rotational speed (Ugender, 2018) [2] as shown in Figs. 7 and 8.

4.2 Flexural Strength

Three-point twisting tests were completed for similar FSW joints. Three-point forces were exerted where one force was exerted from the up side and the other two forces were exerted from the down side. The crossover face and root flexural test were implied for the assessment of malleability of the FSW joints for various tool rotational velocities. Table 2 outlines the decrease in compressive strength and the increase in tool rotational velocity.

The compressive strength of the samples (C-1 to C-3) decreased from 5% to 20% with increasing the tool rotation velocity to 500 rpm as compared to the samples (C-4 to C-6) with rotational speed 1000 rpm as shown in Table 2. It may be due to more heat generated in the welded zone. The compressive strength obtained at samples of rotational speed 500 rpm and samples of rotation speed 1000 rpm shown in Fig. 9. Many researchers have found similar trends on decrease in the compressive strength with increasing rotational speed (Ugender, 2018) [2] as shown in Fig. 9.

4.3 Impact Toughness

The impact test of similar material friction stir welding welded joints was assessed and introduced for two tool rotational velocities. The effect of impact energy of FSW joint with the notch was set at the weld middle line. Table 3 delineates that a rise in the rotational velocity 500 rpm to 1000 rpm increases the impact energy. The impact energy for the tool velocities of 1000 rpm is 5.1 J. It has been observed that material developed at a tool rotation velocity of 1000 rpm possessed higher impact energy 5.1 J.

The impact strength of the samples (C-4 to C-6) increased from 5% to 15% with increasing the tool rotation velocity to 1000 rpm as compared to the samples

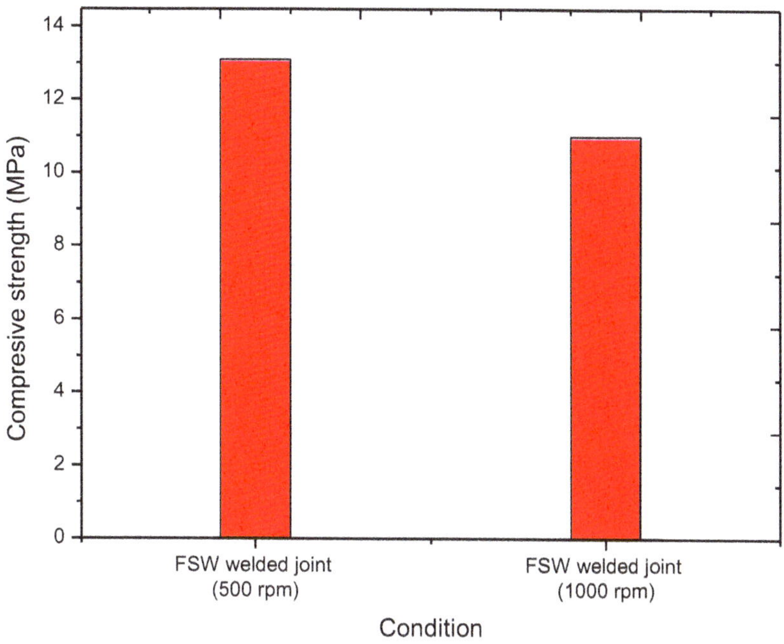

Fig. 9 The impact of the welding rotational speed on the compressive strength in the friction stir welded workpieces

Table 3 Compressive strength in FSW workpieces due to varying rotational velocity

Tool rotation velocity (rpm)	Welding velocity (mm/min)	Impact energy (J)			Avg. impact energy (J)
		Trial 1	Trial 2	Trial 3	
500	28	3.9	4.5	4.7	4.3
1000	28	4.9	5.2	5.4	5.1

(C-1 to C-3) with a rotational speed 500 rpm as shown in Fig. 10. It may be due to more heat generated in the welded zone so that impact strength is high in samples of rotational speed 1000 rpm as compared to the samples of rotation speed 500 rpm as shown in Fig. 10. Many researchers have found similar trends on increase in the impact strength with increasing the rotational speed (Ugender, 2018) [2]. The impact strength of the samples increased with increasing the tool rotational velocity.

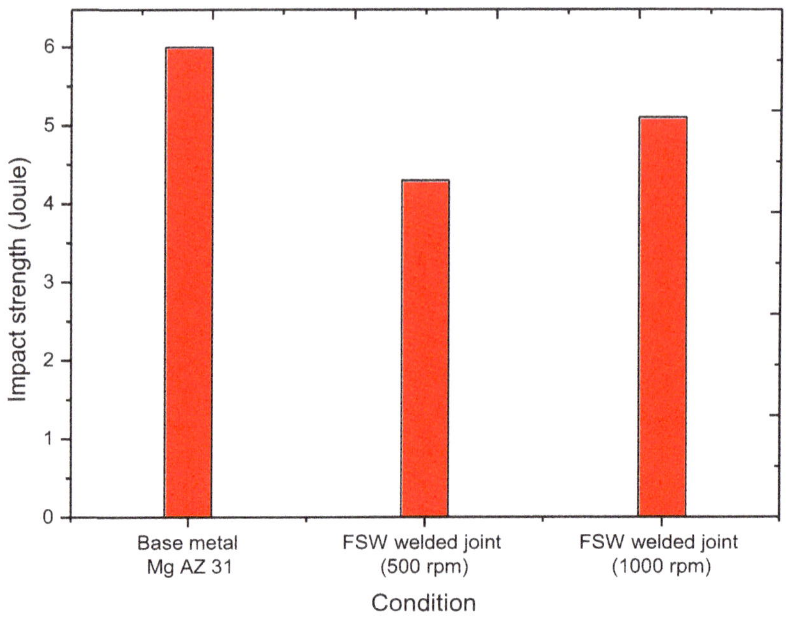

Fig. 10 The impact of the welding rotational speed on the impact strength in the friction stir welded workpieces

5 Conclusions

The tool material and rotational velocity have been recognized as the significant parameters that influence the mechanical properties (tensile strength, impact strength, flexural strength) of FSW measure. The conclusions from the study can be summarized as follows:

1. Magnesium AZ31 alloy was effectively friction stir welding welded joints under the accompanying limit of process parameters: cylindrical tool pin profile, tool revolution velocity of 500–1000 rpm, and welding velocity constant 28 mm/min.
2. The joint manufactured at a rotational velocity 500 rpm and welding velocity 28 mm/min has demonstrated lower UTS and percentage of elongation compared with the joints manufactured with tool rotational velocity 1000 rpm and welding velocity 28 mm/min.
3. The joint manufactured at a rotational velocity 500 rpm and welding velocity 28 mm/min has demonstrated lower impact strength compared with the joints manufactured with the tool rotational velocity 1000 rpm and welding velocity 28 mm/min.
4. The joint manufactured at a rotational velocity 500 rpm and welding velocity 28 mm/min has demonstrated higher compressive strength compared with the

joints manufactured with the tool rotational velocity 1000 rpm and welding velocity 28 mm/min.
5. Tensile properties increase with an increase in the rotational velocity of the tool.
6. Of the two joints manufactured utilizing two distinctive tool rotational velocities, the joint manufactured utilizing with mild steel tool with rotational speed of 1000 rpm and welding velocity 28 mm/min displayed prevalent tensile strength properties.
7. The FSW welded joint shows UTS 203.3 MPa, % of elongation 10.9, and impact energy 5.1 J with tool rotational velocity 1000 rpm and welding velocity 28 mm/min.
8. The FSW welded joint shows compressive strength at a tool rotational velocity 1000 rpm and welding velocity 28 mm/min.
9. FSW is primarily impacted by welding velocity which is trailed by rotational velocity of the tool.

References

1. Singh, K., Singh, G., & Singh, H. (2018). Review on friction stir welding of magnesium alloys. *Journal of Magnesium and Alloys, 6*(4), 399–416.
2. Ko, Y., Lee, K., & Baik, K. (2017). Effect of tool rotational speed on mechanical properties and microstructure of friction stir welding joints within Ti–6Al–4V alloy sheets. *Advances in Mechanical Engineering, 9*(8), 1–7.
3. Liu, D., Xin, R., Zheng, X., Zhou, Z., & Liu, Q. (2013). Microstructure and mechanical properties of friction stir welded dissimilar Mg alloys of ZK60–AZ31. *Materials Science and Engineering: A, 561*, 419–426.
4. Liu, F., Yan, J., & Bai, Y. (2020). Influence of multipass high rotating speed friction stir pro cessing on microstructure evolution, corrosion behavior and mechanical properties of stirred zone on AZ31 alloy. *Transactions of Nonferrous Metals Society of China, 30*(12), 3263–3273.
5. Singh, R. P., Dubey, S., Singh, A., & Kumar, S. (2020). A review paper on friction stir welding process. *Materials Today: Proceedings, 38*(1), 6–11.
6. Singh, V. P., Patel, S. K., Kumar, N., & Kuriachen, B. (2019). Parametric effect on dissimilar friction stir welded steel-magnesium alloys joints: A review. *Science and Technology of Welding and Joining, 24*(8), 653–684.

Correlation of Mechanical and Rheological Properties of Al-Al2O3-Nylon 6 Composite Feedstock Filament for Rapid Tooling

K. S. Boparai and R. Singh

1 Introduction: Background

Presently, fused deposition modelling (FDM) emerges as the second most widely used 3D printing technology for the production of prototypes and functional parts. The technology also has shown promising results in biomedical applications such as dentures, implants, cell printing, etc. The literature reported the development of feedstock filaments of different polymeric composites. Moreover, various researchers have used different blending techniques (such as mechanical blending, chemical blending, chemical-assisted mechanical blending, etc.) for uniform mixing and homogenous composition [1–3]. Basically, the need of alternative feedstock filaments is to widening the application range of FDM [4]. The reinforcement of nano graphene powder in polymeric matrix improves the mechanical properties, also induces piezoelectric property inside the material matrix, and causes a conversion of non-smart matrix into smart matrix [5–7]. Another study reported the reinforcement of barium titanate ($BaTiO_3$) and graphene powder in polyvinylidene fluoride (PVDF) for the enhancement of piezoelectric properties [8].

The developed alternative feedstock filaments show improvement in thermal, mechanical and wear properties. Boparai et al. [2] developed a Nylon 6 based, reinforced with aluminium (Al) and aluminium oxide ($Al2O_3$) feedstock filament, as a substitute of ABS standard feedstock filament. The developed feedstock filament

K. S. Boparai
Department of Mechanical Engineering, MRS Punjab Technical University Bathinda, Bathinda, Punjab, India

R. Singh (✉)
Department of Mechanical Engineering, National Institute of Technical Teachers Training and Research, Chandigarh, India
e-mail: rupindersingh@nitttrchd.ac.in

C. Prakash et al. (eds.), *Additive, Subtractive, and Hybrid Technologies*, Mechanical Engineering Series, https://doi.org/10.1007/978-3-030-99569-0_8

has excellent thermal and wear properties and can be considered for the direct fabrication of thermally stable and wear-resistive prototypes/functional parts. In another study, almond skin powder (food waste) reinforced in PLA for the fabrication of feedstock filament and 3D printing of biomedical implants. Ranjan et al. [9] also made the reinforcement of hydroxyapatite (HAp) and chitosan (CS) in PLA matrix for the preparation of biocompatible and biodegradable feedstock filament, which can be used for various medical applications. Turku et al. [10] prepared the feedstock filament from the waste polystyrene, acrylonitrile butadiene styrene and polyvinylchloride. Osman and Atia [11] reinforced rice straw in ABS matrix for the development of composite feedstock filament for FDM system. Further, the results highlighted that the flexural strength decreases and water absorption decreases with the increase of fibre contents. Prioritization and correlation can be used for the efficient optimization of process parameters [12, 13].

A significant research work has been carried out to explore the potential of alternative feedstock filaments in terms of mechanical, thermal, wear and morphological properties. The literature also supports the significant improvements observed in the above-said properties in comparison with standard feedstock filament. However, little has been reported on the mathematical development of correlation matrix for the qualitative evaluation of various properties of the developed feedstock filament. In the present research work, an effort has been made to develop a correlation among the responses (tensile strength, % elongation at break, young's modulus, MFI and tan δ) of feedstock filament prepared with a Nylon 6 as a base material and reinforced with aluminium (Al) and aluminium oxide (Al_2O_3). The tensile testing and rheological data have been taken from the previous research work done by Boparai et al. [2]. Further, it is a necessary step to predict the exact correlation among variables to thoroughly understand the mechanical performance of developed feedstock filaments.

2 Materials and Methods

As already quoted, Nylon 6 is a polymeric base material, and the reinforced materials are Al and Al_2O_3. The synthesis of the materials and the alternative feedstock filament development process have been explained in previous work [2]. Table 1 shows the weight proportions of the developed compositions. Table 2 reports different sets of variables and their key information for the development of correlation matrix.

Table 1 Weight proportions of the developed composition.

Composition	Nylon 6	Al	Al_2O_3
A	60	26	14
B	60	28	12
C	60	30	10

3 Development of Correlation Matrix and Discussions

A convenient way to represent correlation coefficient among variable is through a correlation matrix. A correlation matrix is a table that shows correlation coefficient between different sets of variables. Table 3 shows the mechanical properties, rheological properties and tan δ values of the alternative feedstock filament and ABS filament. All the specimens have been tested under similar set of condition. The correlation coefficients with a different set of mechanical properties, rheological properties and tan δ values of the alternative feedstock filament and ABS filament have been reported in Table 4.

As per the data available in Table 5 regarding correlation strength, it has been observed that tensile strength has a perfect relation with elongation at break, young's modulus and MFI values. It is little bit less but strong strength with tan δ. Elongation at break has perfect strength with tensile strength, Young's modulus and MFI

Table 2 Different sets of variables and their key information for development of correlation matrix

Variable name	Key information
Tensile testing 1. Tensile strength (TS) 2. Elongation at break (EL) 3. Young's modulus (YM)	Strength of material while in stretch under UTM testing
Melt flow index value (MFI)	Melt flow index of polymeric composite
Tan δ	Viscoelastic property measured by dynamic mechanical analysis

Table 3 Mechanical, rheological and thermal properties of developed alternative feedstock filaments

Composition/material	TS (MPa)	% EL @ Break	YM (MPa)	MFI	Tan δ
A	21.40	18.62	582	2.19	0.11
B	21.53	12.74	760	2.25	0.098
C	21.65	8.56	1165	2.31	0.087
ABS (Acrylonitrile butadiene styrene)	22	6.0	1627	2.41	0.225

Table 4 Correlation matrix

	TS	EL	YM	MFI	Tan δ
TS	1	−0.90441	0.983301	0.989849	0.850174
EL	−0.90441	1	−0.93745	−0.95447	−0.54615
YM	0.983301	−0.93745	1	0.993322	0.767337
MFI	0.989849	−0.95447	0.993322	1	0.766988
tan δ	0.850174	−0.54615	0.767337	0.766988	1

Table 5 Correlation strength

Correlation strength	Positive	Negative
Perfect	0.9 to 1	−0.9 to −1
Strong	0.5 to 0.9	−0.5 to −0.9
Weak	0.1 to 0.5	−0.1 to −0.5
Uncorrelated	0 to 0.1	0 to −0.1

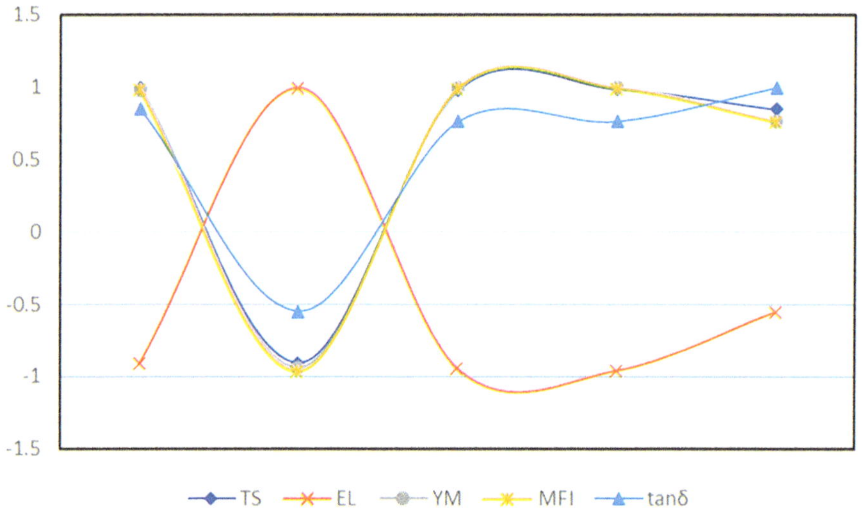

Fig. 1 Correlation graph

values; however, it has stronger strength with tan δ. In the case of Young's modulus, it has a perfect relation with tensile strength, elongation at break and MFI but stronger relation with tan δ. On the other hand, MFI values also show similar trend. Finally, tan δ has a strong relation with all the responses.

Figure 1 illustrates the graphical representation of the variables.

4 Conclusions

The present work developed different correlation matrix among different variables of feedstock filament development on lab scale, and from the study following points, it may be concluded:

1. Tan δ of materials has shown least correlation (as compared to other responses) with other variables, thus it may be concluded that Tan δ is inherent property of material and is least bothered with other processing variables.
2. Tensile strength, elongation at break, Young's modulus and MFI values have perfect correlation with each other.

Acknowledgement The authors are extremely thankful to MRSPTU, Bathinda and NITTTR Chandigarh for technical support.

References

1. Beesetty, P., Patil, B., & Doddamani, M. (2020). Mechanical behavior of additively manufactured nanoclay/HDPE nanocomposites. *Composite Structures, 247*, 112442.
2. Boparai, K. S., Singh, R., & Singh, H. (2016). Experimental investigations for development of Nylon6-Al-Al2O3 alternative FDM filament. *Rapid Prototyping Journal, 22*(2), 217–224.
3. Sandhu, G. S., & Singh, R. (2019). Development of ABS-graphene blended feedstock filament for FDM process. In *Additive manufacturing of emerging materials* (pp. 279–297). Springer.
4. Momeni, F., Liu, X., & Ni, J. (2017). A review of 4D printing. *Materials & Design, 122*, 42–79.
5. Camargo, J. C., Machado, Á. R., Almeida, E. C., & Silva, E. F. M. S. (2019). Mechanical properties of PLA-graphene filament for FDM 3D printing. *The International Journal of Advanced Manufacturing Technology, 103*(5), 2423–2443.
6. Ivanov, E., Kotsilkova, R., Xia, H., Chen, Y., Donato, R. K., Donato, K., Godoy, A. P., Di Maio, R., Silvestre, C., Cimmino, S., & Angelov, V. (2019). PLA/graphene/MWCNT composites with improved electrical and thermal properties suitable for FDM 3D printing applications. *Applied Sciences, 9*(6), 1209.
7. Yao, T., Ye, J., Deng, Z., Zhang, K., Ma, Y., & Ouyang, H. (2020). Tensile failure strength and separation angle of FDM 3D printing PLA material: Experimental and theoretical analyses. *Composites Part B: Engineering, 188*, 107894.
8. Kim, H., Fernando, T., Li, M., Lin, Y., & Tseng, T. L. B. (2018). Fabrication and characterization of 3D printed BaTiO3/PVDF nanocomposites. *Journal of Composite Materials, 52*(2), 197–206.
9. Ranjan, N., Singh, R., Ahuja, I. S., & Singh, J. (2019). Fabrication of PLA-HAp-CS based biocompatible and biodegradable feedstock filament using twin screw extrusion. In *Additive manufacturing of emerging materials* (pp. 325–345). Springer.
10. Turku, I., Kasala, S., & Kärki, T. (2018). Characterization of polystyrene wastes as potential extruded feedstock filament for 3D printing. *Recycling, 3*(4), 57.
11. Osman, M. A., & Atia, M. R. (2018). Investigation of ABS-rice straw composite feedstock filament for FDM. *Rapid Prototyping Journal.*
12. Amin, M. A., & Karim, M. A. (2013). A time-based quantitative approach for selecting lean strategies for manufacturing organisations. *International Journal of Production Research, 51*(4), 1146–1167.
13. Krol, T. A., Seidel, C., & Zaeh, M. F. (2013). Prioritization of process parameters for an efficient optimisation of additive manufacturing by means of a finite element method. *Procedia CIRP, 12*, 169–174.
14. Singh, R., Kumar, R., Pawanpreet, Singh, M., & Singh, J. (2019). On mechanical, thermal and morphological investigations of almond skin powder-reinforced polylactic acid feedstock filament. *Journal of Thermoplastic Composite Materials*. https://doi.org/10.1177/0892705719886010

A Framework on Electrochemical Machining of ABS-15% Al Composite

Arun Dutt Sharma and Rupinder Singh

1 Introduction

The ECM is one of the unconventional machining processes, which is being widely used for machining of electrically conducting surfaces. The surface finish obtained from this machining method is finest among all the unconventional machining methods because the material is removed from the conductive worksurface at the atomic level due to the flow of high current and relatively low voltage with the help of an electrolyte which is generally a water-based salt solution of sodium chloride or sodium nitrate [1]. Kozek et al. [2] presented the relationship between the shape dimensions transferred on the workpiece with the help of tool electrode. The ultra-short pulse current and ultra-small gap size for enhancing machining capabilities of ECM process has been recommended in this work. Also, the use of non-profiled electrode and electrochemical coping of slots, insulating grooves and mini holes was presented. Davydov et al. [1] discussed the theory of electrochemical shaping. The goal was to produce workpiece of specified shape and size. Klocke et al. [3] reported the electrochemical machinability of the selected modern and Ti-Ni-based alloy for aerospace sector. The surface properties were examined with the help of SEM and EDS analyses. da Siva Neto et al. [4] studied the analysis of material removal rate (MRR), analysis of roughness and analysis of lateral over cut and concluded that the feed rate was the important parameter that affects the MRR most, and sodium nitrate shows the better results among all the electrolyte when dealt with roughness and overcut. Kim et al. [5] introduced a disk-type electrode to eliminate the taper. In this study, the authors used a 10 μm diameter wire as a tool electrode for ECM of stainless steel substrate. Hyness and Kumar [6] reported

A. D. Sharma (✉) · R. Singh
Department of Mechanical Engineering, National Institute of Technical Teachers Training and Research, Chandigarh, India

C. Prakash et al. (eds.), *Additive, Subtractive, and Hybrid Technologies*,
Mechanical Engineering Series, https://doi.org/10.1007/978-3-030-99569-0_9

experimental investigations on ECM of Al matrix composites such as SiC reinforced Al, B_4C reinforced Al matrix composites. Used ECM for Al–SiC composite matrix with varying quantities of SiC and with different grades of Al and reported the different optimised process parameters of ECM for MRR, surface roughness (SR), radial overcut (ROC). Analysed the performance of ECM with different grades of aluminium reinforced with boron carbide (B_4C) composite matrix and studied the MRR, SR, ROC of the composite matrix. Formed a non-circular hole with a tool electrode having a circular profile with the help of ECM to meet the aeronautical, marine and industrial applications. Fan and Hourng [7] used 510 μm tungsten rod as anode to fabricate the micro pin which is used in scanning tunnelling microscope. Nickel plate was used as cathode. The authors reported that to fabricate microelectrode with diameter less than 100 μm, low applied voltage with high concentration of electrolyte and appropriate rotation of microelectrode is required. Mathew and Sundaram [8] proposed the ECM of micro tools to overcome the surface cracks, residual stress and deformation which came in existing micro tool fabrication processes. In this work a mathematical was developed to predict the diameter of micro tool fabricated. Soni and Thomas [9] presented a comparative study for drilling of LM6 Al/B_4C composite after obtaining optimized results from Taguchi method and genetic algorithm. Fang et al. [10] introduced a pulsating electrolyte flow which was used to improve the heat transfer and MRR in ECM. Experiments were conducted to verify a multi-physic model coupling of electric, heat, transport of dilute species and fluid flow. Loherngel et al. [11] used a special flow through capillary microcell which controlled the anodic dissolution at high current densities (100A/cm^2) with potentiostatic control in a three-electrode arrangement. Egashira et al. [12] reported the successful drilling of micro holes of diameter 4.5 μm by using ECM and investigate the ECM conditions for high-speed drilling without widening the gap. Saravanan et al. [13] reported a study of ECM of Al metal matrix (AA6603) reinforced with TiC (0–9 wt%). Surface roughness was measured by varying the percentage of reinforced, and best surface finish was observed at 6% by weight of TiC reinforcement. Paczkowgki and Zdrojewski [14] presented system for controlling and monitoring the electrochemical machining of complex surfaces. The parameters were modified according to the designed criteria. Sathish [15] reported the optimization of micro drilling process which was carried out for stainless steel. The study was evaluated to observe the effect of response and duty ratio parameters such as machining time, overcut and MRR; conicity was analysed using VMS images. Hu et al. [16] proposed a new method of gas film insulation to decrease the stray corrosion with the help of compressed air in forward flow mode which was supplied into the cathode and through the workpiece around the machined worksurface to remove stray electrolyte. Kumar et al. [17] performed the ECM of titanium grade 5 which has very good applications in medical and aerospace industries. From experimental work, authors reported that electrolyte concentration with voltage and frequency plays an important role in MRR, and acidic nature of electrolyte affects the MRR. So, in recent years, consequential work has been done on parametric optimization of different process parameters using different substrate material. From the study and mathematical modelling reported in

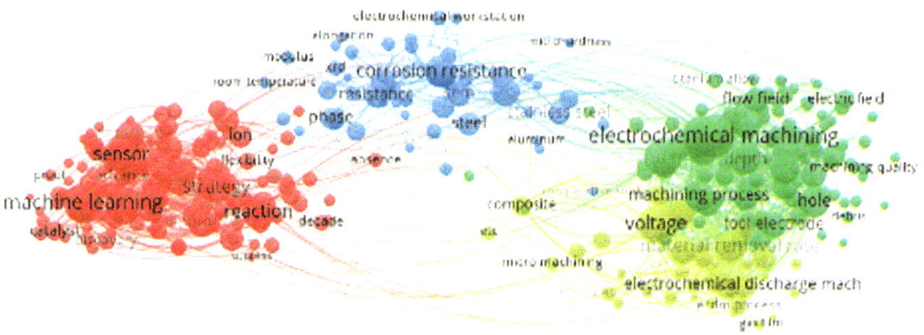

Fig. 1 Bibliographic relevance analysis of ECM by VOS viewer

literature, it is concluded that the MRR, surface finish, surface hardness and other output parameters depend mainly on some input parameters such as electrochemical equivalence, feed rate, temperature, pulse frequency, current and voltage, electrode gap, material density, machining time, electrolyte concentration. Based on Web of Science data, the literature on ECM has been explored for past 20 years (Fig. 1). Figure 2 shows the research gap which clearly highlights that no work has been reported on injection moulded ABS-15% Al for ECM. To further refine the research gap the keyword as ECM of thermoplastic composite reinforced with metal powder was put in web of science databased followed by processing in VOS viewer software, it has been ascertained that no work has been reported on ECM of non-conducting substrate material.

2 Research Approach

The whole study has been conducted in three stages. Figure 3 shows the flowchart for stage 1. In stage 1, it has been observed that no weight loss or material removal rate was noticed with thermoplastic during ECM for the selected input process parameters. Because ABS is a nonconductive material, it was made conductive by reinforcing Al for ECM process. The composite was prepared with the help of injection moulding process, after that electrical conductivity was checked with the measurement of dielectric constant (at 20 °C and 1 kHz frequency); based on dielectric constant value, it has been ascertained that composite material has conductivity after reinforcement of Al. Experiments were performed with a combination of two input process parameters (tool feed rate and voltage), and it was observed that 20–30 mg of the material was removed which was nonsignificant, and further, it was observed that no tool profile was generated on the workpiece. The SEM and EDS analyses were performed on the selected sample to check the composition of ABS-15% Al before and after ECM. In stage 1, experiment with ECM was performed on the aluminium substrate without slit for process parametric optimization. Further in stage 3, ECM may be performed with optimized process parameters

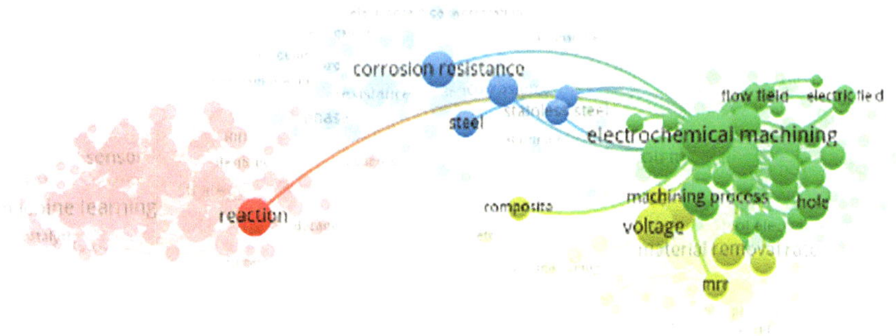

Fig. 2 Bibliographic relevance analysis of the ECM

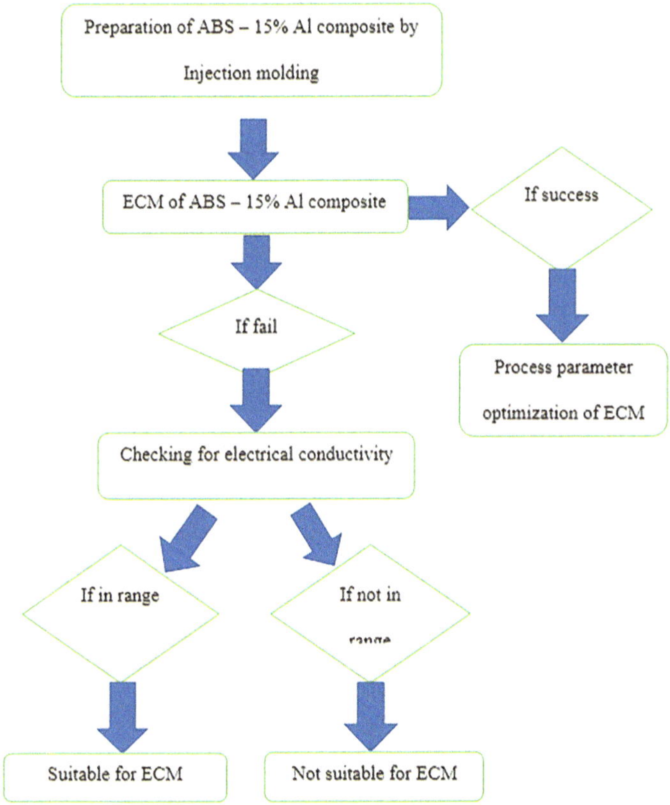

Fig. 3 Flowchart for stage 1

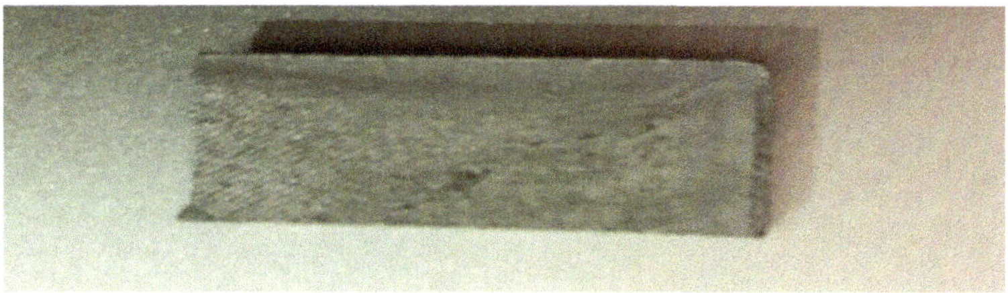

Fig. 4 Injection moulded slit of ABS-15% Al

Fig. 5 Slit having circular hole

Fig. 6 Slit having rectangular hole

which is ascertained from stage 2 with slit of having same cross-sectional area of different shapes (Figs. 4, 5, and 6).

3 Results and Discussion

At the first stage when ECM was performed on ABS-15% Al with electrolyte (NaOH) 100 g/l, feed rate of 0.144 mm/min, at 25volt, using 4 mm tool and work-piece gap for machining time of 1800s, some material removal was observed. Figures 7 and 8 show SEM and EDS analyses before and after ECM. As observed from Figs. 7 and 8, there is a significant variation of element present in ABS–15% Al. It was observed that Al percentage was reduced after ECM in the form of weight loss.

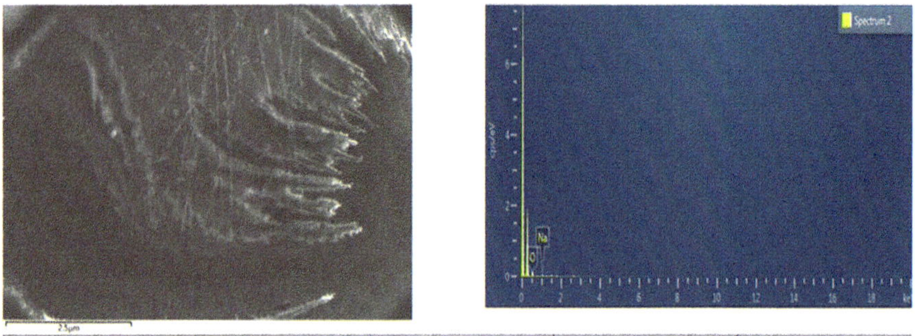

Spectrum 2				
Element	Line Type	Weight %	Weight % Sigma	Atomic %
C	K series	79.53	1.38	84.32
O	K series	18.54	1.38	14.76
Cl	K series	0.54	0.11	0.19
Na	K series	0.85	0.19	0.47
Al	K series	0.54	0.12	0.25
Total		100.00		100.00

Fig. 7 SEM and EDS report before ECM

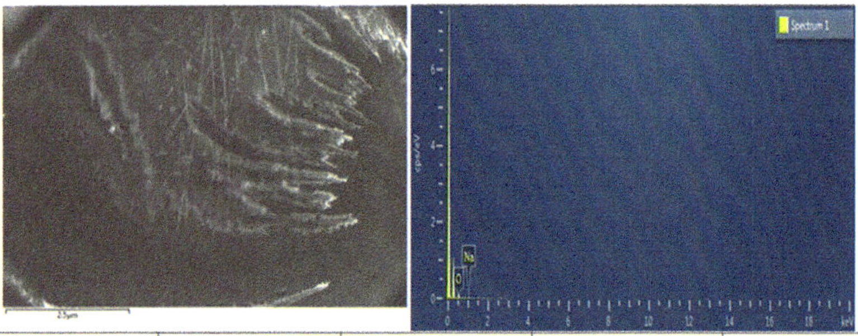

Spectrum 1				
Element	Line Type	Weight %	Weight % Sigma	Atomic %
C	K series	74.78	1.88	80.28
O	K series	23.39	1.89	18.85
Cl	K series	0.58	0.13	0.21
Na	K series	0.83	0.23	0.47
Al	K series	0.42	0.13	0.20
Total		100.00		100.00

Fig. 8 SEM and EDS report after ECM

4 Conclusions

An experimental study was performed on ECM of ABS-15% Al thermoplastic composite. In this work, a framework was presented for ECM of ABS which was reinforced with 15% Al through injection moulding. Following are the conclusions of this study:

- The ECM of thermoplastic matrix can be possible with reinforcement of some metal powder, but significant metal removal rate was not observed that's why process parametric optimization and mathematical relation cannot be performed at the first stage.

- Because material removal rate was not significant, so there is a huge possibility to use the slit as the tool insert and to guide the electrolyte while ECM.

References

1. Davydov, A. D., Volgin, V. M., & Lyubimov, V. V. (2004). ECM of metals: Fundamental of electrochemical shaping. *Russian Journal of Electrochemistry, 40*(12), 1230–1265.
2. Kozek, J. R., Kamlakar, P., & Makker, Y. (2004). Selected problem of micro – Electctrochemical machining. *Journal of Materials Processing Technology, 149*(2004), 426–431.
3. Klocke, F., Zeis, M., Klink, A., & Veselovac, D. (2013). *Experimental research on the ECM of modern titanium and nickel-based alloys for aero engine components*. Elsevier.
4. da Siva Neto, J. C., da Silva, E. M., & da Silva, M. B. (2006). Intervening variables in ECM. *Journal of Material Processing Technology, 179*, 92–96.
5. Kim, B. H., Na, C. W., Lee, Y. S., Choi, D. K., & Chu, C. N. (2005). Micro ECM of 3D micro structure using dilute sulphuric acid. *CIRP Annals, 54*(1), 191–194.
6. Hyness, R. J., & Kumar, R. (2018). ECM of aluminium metal matrix composites. *Surface Engineering and Applied Electrochemistry, 54*(4), 367–373.
7. Fan, Z.-W., & Hourng, L.-W. (2009). The analysis and investigation on the microelectrode fabrication by ECM. *International Journal of Machin Tool and Manufacturer, 49*, 659–666.
8. Mathew, R., & Sundaram, M. (2012). Modelling and fabrication of micro tools by pulsed ECM. *Journal of Material Processing Technology, 212*(7), 1567–1572.
9. Soni, S. K., & Thomas, B. (2017). A comparative study of ECM process parametric by using GA and Taguchi method. *Materials Science and Engineering, 263*(2017), 062038.
10. Fang, X. L., Qu, N., Zhang, Y., Xu, Z., & Zhu, D. (2014). Effect of pulsating electrolyte flow in ECM. *Journal of Material Processing Technology, 212*(1), 36–43.
11. Loherngel, M. M., Rataj, K. P., & Munningoff, T. (2016). ECM mechanisms of anodic dissolution. *Eclectrochimica Acta, 201*, 348–353.
12. Egashira, K., Hayashi, A., Hirai, Y., Yamaguchi, K., & Ota, M. (2018). Drilling of microholes using ECM. *Precision Engineering, 54*, 338–343.
13. Saravanan, S., Balan, A. V., Dinesh, S., & Vijayan, V. (2020). ECM behavior of AA6063 – TiC composites by using response surface methodology. *Material Proceeding, 21*(Pt 1), 592–594.
14. Paczkowgki, T., & Zdrojewski, J. (2017). Monitoring and control of the ECM process under the condition of a vibrating tool electrode. *Journal of Material Processing Technology, 244*, 204–214.
15. Sathish, T. (2019). Experimental investigation of machined hole and optimization of machining parameters using ECM. *Journal of Material Research and Technology, 8*(5), 4354–4363.
16. Hu, X., Zhu, D., Li, J., & Gu, Z. (2019). Flow field research on ECM with gas film insulation. *Journal of Materials Processing Technology, 267*, 247–256.
17. Kumar, V. P., Soundrapandian, E., Tajdeen, A., & Prashanth, T. (2021). Experimental study of ECM of titanium (Ti-6Al-4V) alloy. In *Advances in materials research* (pp. 301–308).

Cost-Effective Design of Soft Robotic Prosthetic Arm Based on 3D Printing

Abhijeet Singh, Sukhinderpal Singh, and Raman Kumar

1 Introduction

Human hand is a multifaceted edifice and is robust but precise. It comprises a multipart mechanical construction that permits the hand to familiarize and competently handle numerous shapes, weights and textures. Nowadays, prosthetic devices cannot deliver various functions and become complicated and expensive [1]. The prosthetic devices are self-possessed of several specific parts: miniaturized actuators, sensors and complex control. These collectively satisfy the manipulation tasks performed by a prosthetic device. There is a significant need for a reasonable and straightforward prosthetic hand to help amputees in countries like India. Various studies reported to deal with these problems, and some considerable researches are documented.

The concept for a robotic hand prosthetic fabricated via Shape Deposition Manufacturing (SDM) was developed and discussed. The hand was reliable and durable and incorporates a simple design that requires only a single actuator for the eight active degrees of freedom [2]. A prosthetic arm developed using pneumatic actuators and a tendon-driven wrist using a wire drive. The prosthetic hand was safe when it contacted people because of the pneumatic actuators [3]. Dalley et al. [4] presented an anthropomorphic hand's design and performance data for trans-radial amputees. The hand described was capable of eight canonical postures. A tendon-driven robot hand was developed and proposed with a pressure-dependent finger stiffness and described the driving mechanism, primary structure, and properties of the offered robot hand [5]. An attempt to understand the hysteresis peculiar to the muscle was presented by a detailed review of the theory of McKibben artificial

A. Singh · S. Singh · R. Kumar (✉)
Department of Mechanical Engineering, Guru Nanak Dev Engineering College, Ludhiana, Punjab, India

C. Prakash et al. (eds.), *Additive, Subtractive, and Hybrid Technologies*, Mechanical Engineering Series, https://doi.org/10.1007/978-3-030-99569-0_10

pneumatic muscles [6]. Sun et al. [7] studied the bending and rotary motion of soft pneumatic actuators. The measured force outputs of different actuators were analysed and presented. A physical model was introduced to give insight into the behaviour of the actuators. The design and fabrication of a fibre-reinforced soft pneumatic actuator demonstrated, whose bending axis and bend radius could be mechanically programmed with a flexible, conformal covering that constrains the motion achieves greater gripping force mechanically [8]. A soft robotic thumb rehabilitation system is presented as a task-oriented therapeutic device that could restore and replicate exact thumb motion for patients with neurological hand disabilities [9]. A low-cost prosthetic arm for trans-radial amputees was proposed. The arm was entirely 3D printed with Arduino Mega as a controller complaining of an electromyography (EMG) sensor. The actuation force was limiting and provided input for necessary mechano-tactile haptic feedback [10]. A design based on a pneumatic system was described to operate at low flow rates and thus increases the actuation speed. An actuator could bend from a normal to a quasi-circular shape in about 50 ms when pressurized at 345 kPa. A comparably lesser strain was observed in a material at complete actuation [11].

The effect of fibre angle on fibre-reinforced soft pneumatic actuators' deformation was investigated. While they extend axially, the behaviour expands radially and twist about their axis as a function of inlet pressure was also studied [12]. A hand with fully actuated fingers and a single degree of freedom wrist were reported. Pneumatic muscles were used for the functioning of the biceps and triceps. The prosthetic hand showed a 250 g of grip strength with a similar speed and movement of a human hand [13]. A new three-dimensional printed prosthesis developed for patients with trans-radial limb amputation powered with an anthropomorphic terminal device. The results reveal that the independent thumb movement facilitated object grasp with a low weight [14]. Beyrouthy et al. [15] proposed designing electroencephalography (EEG) controlled, smart, 3D printed prosthetic arm. It has smart sensors and actuators to give feedback about the environment and the object in contact. The workability of 3D printable technology was explored to develop low-cost robotic prosthetics [16]. An inexpensive 3D printed mechanical shoulder prosthesis was created to help a child perform bi-manual activities and achieve improved balance and performance of some bimanual activities [17]. A design algorithm is presented to optimize the design parameters for an actuator that will follow an input kinematic trajectory upon pressurization by utilizing analytical modelling based on nonlinear elasticity and optimization. The designed actuators perform the thumb and index finger [18]. A cost-effective prosthetic hand for a patient was presented who lost both hands due to a work accident. Design concepts of the prosthetic hand were successfully developed and tested [19]. Asymmetric bellow flexible pneumatic actuator was investigated as a bending joint made of an elastic material to prototype a multi-jointed, multi-fingered soft robotic hand [20]. A human-like robotic arm named 'Reachy' was designed with seven joints from shoulder to wrist. The structure being 3D printed makes it low cost and based on an open-source architecture [21]. A low-cost 3D printed prosthetic hand was designed and investigated, which is light in weight and promotes active residual limb cooling. EMG signals were acquired from the residual limb muscles to describe hand and wrist activity [22]. A low-cost 3D printed prosthetic arm with soft grippers was presented. The results showed average effectiveness of 87% in grasping and demonstrating soft grippers' efficacy as an end device in prosthetics [23]. A 3D printed prosthetic hand prototype capable of performing four grip patterns with an average response time of 300 ms was controlled wirelessly using EMG [24]. An alternative approach for the fabrication of soft pneumatic fibre-reinforced actuators was proposed. It reduced the labour and time for the manufacturing of actuators and required only a desktop 3D printer [25].

1.1 Need and Significance of the Prosthetic Arm Design

Many devices that provide a human hand morphology have been proposed. Very few offer a real human-like shape and appearance; this is considered a severe issue when the device is used as a prosthetic system. So, in the prosthetic arm's proposed design, efforts are made to prepare an ergonomic and aesthetic design. The proposed plan is comparatively affordable, easy to fabricate, and safe for human interaction. This artificial soft prosthetic hand is powered pneumatically that adapts passively to the objects handled due to mechanical compliance. It is pressure-driven and enables three independent fingers and thumb motions for dexterity or in groups when a synergistic finger movement is needed. The hand has a truly anatomical shape, and it is easy to replace and cheap in production.

The actuation method employed is purely pneumatic controlled by the electronics part of the project, which comprises sensors and pumps. Such networks using pneumatics are also called 'pneu-net' that are nothing but small channels in elastomeric materials. The silicone rubber was chosen for producing sophisticated motions with simple controls. The most important task for its operation is to convert our soft actuator's linear motion into a finger's bending movement under the desired pressure generated by the 'pneu-nets'. Also, there is a significant need for the fingers' rapid operation to lag between actuation and bending of fingers. It must not be heavy as eventually it will create discomfort for the amputee. Soft robotic is the most promising solution for all mentioned problems where a comfortable and less expensive prosthesis is possible. In recent years, some new prosthetic hands based on 3D printing technology have been proposed that make the fabrication of hand relatively easy by creating moulds with required dimensional accuracy and tolerances.

The bio-mimicked human hand achieves a genuinely anatomical shape and appearance of the human hand. It consists of three degrees of freedom actuation that can be operated in groups for simplicity or controlled separately for an enhanced manipulation precision. Index finger, thumb and remaining fingers collectively have separate valves to achieve bending. The shape, configuration and size of the hand can be easily modified to meet each patient's preferences. The manufacturing cost is less as it makes use of 3D printed moulds and silicone rubbers. Such a property makes it affordable and suitable for children amputees that require frequent change of the device due to their body growth. Since it is made up of soft materials, it is also considered safer than traditional prosthesis.

1.2 Problem Statement

Limbs for Life Foundation's data says, 'Every week 2,996 people lose a limb in the US'. According to the Brown Alpert Medical School, the reasons for amputation include cardiovascular disease, traumatic accidents, infection, tumours, nerve injury (trophic ulceration) and congenital anomalies. The most frequent causes of upper limb amputation are trauma and cancer, followed by vascular complications of disease right arm more frequently involved in work-related injuries. Congenital upper limb deficiency has an incidence of approximately 4.1 per 10,000 live births [26]. The data in Table 1 represent various causes for the need for trans-radial amputation.

Census 2011 reveals that out of India's 1.21 billion population, 26.8 million, i.e., 2.21% of the total population, are disabled either mentally or physically. Out of these, 20% are disabled in movement. Seven persons in every 10 with upper limb amputations have amputations distal to the elbow, with trans-radial amputations being 60% of the total wrist and hand amputations. A person loses a limb, facing staggering emotional and financial lifestyle changes. Prosthetic devices can restore necessary skills and independence but are not available to all amputees at low cost and with high personal customization [27].

1.3 Present Study Design Objective

- The present design aims to develop a wearable device that a trans-radial amputee can use to aid in his/her daily tasks.
- The prime objective is to design a simple, biocompatible, easy to prepare and economically affordable prosthetic arm driven by utilizing pneumatic networks.
- The proposed design is light in weight, easy to handle and a human-safe prosthetic arm. It could serve an amputee without being an economic burden on him/her.

Table 1 Causes for trans-radial amputation and their fractions

Cause	Fraction (%)
Congenital	8.9
Tumour	8.2
Disease	5.8
Trauma	77

2 Methodology

The project is about designing and prototyping a robotic prosthetic arm that utilizes additive manufacturing techniques to prepare its prototype. Being functional for a trans-radial amputee and helping in his daily tasks of picking, holding, gripping, etc. The actuation of the arm is achieved from pneumatically operated artificial muscles called 'pneu-nets'. Its control is achieved by employing Arduino Uno with an electromyography sensor.

The arm's design is prepared in such a way that it can be 3D printed to fabricate and in such a way that it can provide modularity in shape and size to facilitate different patients' requirements. Figure 1 shows a research methodology employed for the project. The very first step in this is to validate the need for this work. This was done partly through a literature survey and partly through online data from various foundations, as mentioned in the problem statement.After a thorough look at the previous work done in this field, visualization was made to address the problem to fulfil the objective simultaneously, and Fig. 2 shows a basic sketch for the design.

Fig. 1 Research methodology to design robotic prosthesis arm

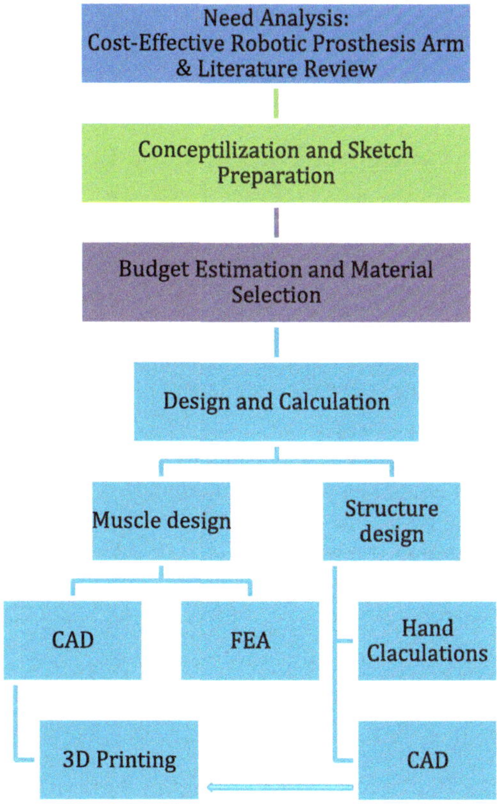

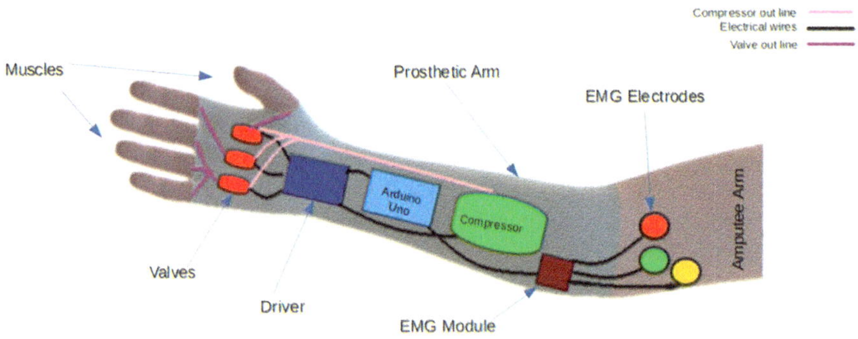

Fig. 2 Diagram of the arm with various components

To meet the aim of developing a low-cost or easily affordable arm, a tight budget was prepared to keep the cost low at the very first step. The proposed material needs to be 3D printed, so according to Acrylonitrile-Butadiene-Styrene (ABS), it was considered for the whole arm structure.The designs for artificial muscle and the arm structure were carried out separately. Muscles being moulded from liquid silicone needed 3D printed moulds designed for a known trajectory, as mentioned later in this case study. The arm's structure was a replica of a human arm and was manipulated to fit all the required components within.

3 Conceptualization and a Basic Idea

For actuation, 'pneu-nets' are employed as artificial muscles and are made from silicone rubber with a shore hardness of 28 A and 5 A. These muscles are used to function as fingers and thumb, driven via a micro reciprocating compressor. Air from the compressor reaches the miniature solenoid valves and from there conditionally to the muscles. Arduino Uno completely processes all the control and motion in coherence with the EMG module and motor driver. The raw signals are sensed from the amputee's bicep using three EMG electrodes and pre-processed in the EMG chipset. These processed signals are then fed into Arduino analogue input, where new signals are processed for programmed output. These signals are then received as an input by the motor driver, which then powers the compressor and solenoid valves accordingly. Figure 2 gives a necessary schematic placement and interconnection of components being used.

3.1 Material Selection

The arm's structure is the primary concern of the design because it supports the complete assembly firmly. At the same time, it must be strong enough to withstand mishandling up to some extent. Taking all these parameters into account, ABS was found to be suitable and could be 3D printed with ease during fabrication. The same material could be found useful for 3D printing the moulds for artificial muscles.

3.2 Design

The hand contains three independent and two dependent pressure-driven soft, flexible actuators.Each artificial muscle serves the purpose of either a finger or a thumb. The hand is designed to be easily manufactured without the use of any expensive equipment. The muscles are made up of three components silicone, strain limiting layer and polyester thread. The elastic property of silicone rubber serves as a bladder that expands on the application of pressure inside. This expansion is being restrained by strain limiting layer and polyester thread. The actuator is made up of two layers of silicone rubber.

3.3 Muscle Design

The actuator is made up of two layers of silicone rubber. The inner part is made with silicone with a shore hardness of A28 and is stiff to give the shape and strength. The cross-section of the actuator is semicircular with adequate wall thickness and is moulded with silicone rubber. Then there is an in-extensible, strain-limiting layer with a fibreglass fabric attached to the actuator core's flat face. This strain-limiting layer prevents the flat face from expanding when the actuator is inflated, resulting in the actuator's curling trajectory. Next, the actuator is wrapped throughout its length with an in-extensible polyester thread. This restricts its radial expansion. This wrapped thread is secured in place by moulding silicone's outer coat with a shore hardness of A5, which encapsulates the threads and the actuator core.

At last, the actuator is capped at both the ends, hence forming a closed chamber. A vented screw is also installed at one of the ends to provide a passage for air to flow in and out. The exploded view of the actuator is shown in Fig. 3.

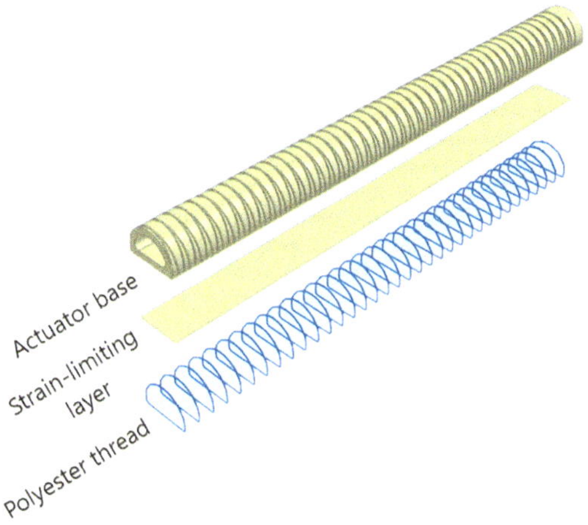

Fig. 3 Assembly of the actuator

3.4 Structure Design

The structure of the arm and palm was designed to be 3D printable. The model is such that it can house various components inside while maintaining suitable strength and enhancing the modularity of design. A 3D model for a human arm was manipulated to accommodate multiple members. Hand calculations were carried out to validate the material's suitable strength and find a rough estimate of thickness to start with.

3.5 Control and Electronics

In order to control the thumb and fingers, Arduino Uno was used with an AD8232 EMG Module. The muscle movement signals are sensed through the electrodes (Medico ECG electrodes) and then transferred to the EMG chip, which converts this raw signal to an analogue form, which Arduino can understand. The connections are shown in Fig. 4. When Arduino receives the signal, a set of programs are executed. The signal is sent from Arduino to the motor driver to control the artificial muscles, which drive the solenoid valves and the compressor. The technical specifications of the compressor are shown in Table 2.

The Arduino code to fetch the output from the EMG module is shown in Fig. 5.

According to the requirement, the Camozzi-Series K8 directly operated solenoid valves (K8000-305-K22) were appropriate for our use through research and suitability. Three such valves were used. These have three ports and two actuation positions. The first port is connected with the muscle, the second to the pump line, and the third is for the exhaust. Table 3 gives general specifications of the valve with dimensions.

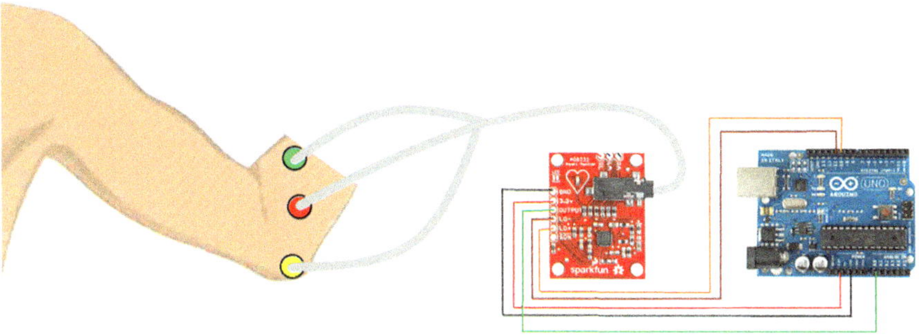

Fig. 4 Wiring diagram for the EMG module

Table 2 Technical specifications of the compressor

Parameters	Range
DC voltage	6–12 V
Working current	1.5 A
Flow rate	1.5 lpm
Maximum pressure	5.5 bar
Noise level	55 dB

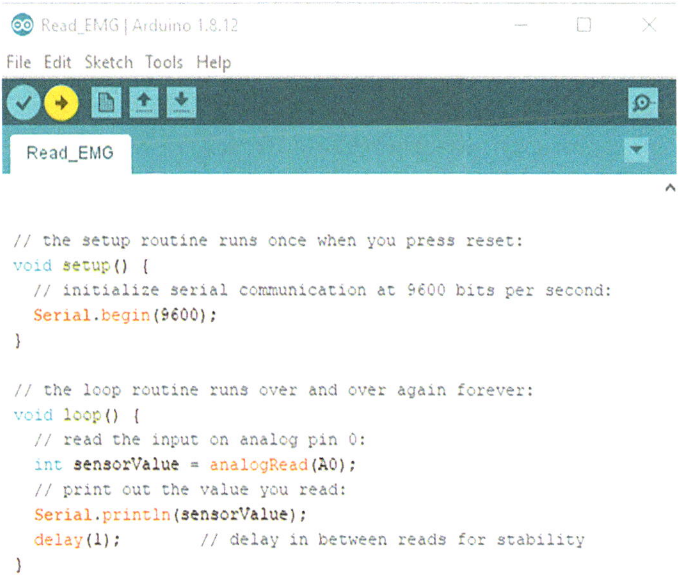

Fig. 5 Arduino Uno code for EMG module

Table 3 Specifications of solenoid valve with dimensions

Sr. No.	Parameters	Range
1	Function	3/2 NO
2	Operation	Poppet type
3	Nominal diameter	0.7 mm
4	Operating pressure	0–7 bar
5	Response time	<10 ms
6	Voltage	12 V DC
7	Power	0.6 W
8	Voltage tolerance	±10%

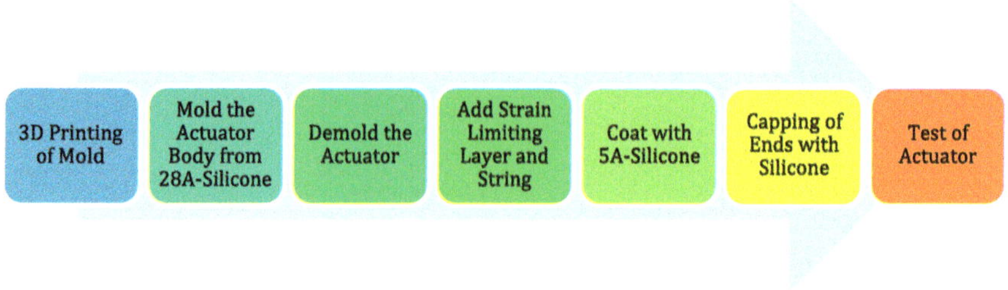

Fig. 6 Fabrication process of artificial muscles actuator

3.6 Actuator Fabrication Process

The fabrication process of an artificial muscles actuator is shown in Fig. 6. It consists of several moulding steps and a set of 3D printed mould.

The process starts by moulding the actuator body (a semilunar tube) out of 28 A-Silicone. After the silicone is set and cures, it is taken out, and a strain-limiting layer is glued. The polyester thread is then wrapped and secured by coating the assembly with silicone 5 A. The one end of the actuator is closed using silicone 28 A, and on another end, a vented screw is plugged and sealed with silicone glue. Now attach the actuator to the compressor and pressurize it to make it bend.

3.7 Utilization of Additive Manufacturing

The process of fabrication of artificial pneumatic muscles requires high precision at small dimensions. In order to achieve such a high manufacturing requirement, the fused deposition modelling (FDM) technique of additive manufacturing was employed. For the moulding of liquid silicone, the moulds were first designed in SolidWorks, sliced in Cura, and then 3D printed in Ultimaker 2+. The filament material being Acrylonitrile-Butadiene-Styrene (ABS) with a diameter of 2.5 mm. Shown in Fig. 7 is a sample 3D printed mould for artificial muscles.

Fig. 7 3D Printed mould for artificial muscle

4 Conclusion

The present work showed that it is feasible to design a soft-robotic prosthetic arm using 3D printing technology, which may mimic a trans-radial arm prosthesis with a low budget, lesser labour and proper operation. The EMG was tested and worked to control the actuation. The structural design is aesthetic and ergonomic for the patient with good strength and functionality. It needs to be mentioned that the most promising advantage of employing 3D printing is that it does not require highly trained professionals, and the arm could be made with high agility, allowing adaptability for different amputees. The actuators could be made more predictable and precise. The whole pneumatic system could be optimized for better performance and lower power requirements. To make the design more compact, electronic controllers could be developed for this particular requirement. The use of a single EMG sensor provided the necessary input for the control of fingers and thumb. However, the use of an Armband from MyoWare could be more accurate and easier to use. Leap motion is another better technology that can be promising for mere testing. The future work could be to make the device more adaptable and compact for the patient.

Further, improvement can be achieved by employing gesture control in the EMG sensor. The use of feedback sensors is also advised to aid in proper control and actuation of the arm while dealing with various objects.

References

1. Fras, J., & Althoefer, K. (2018, October 1–5). *Soft Biomimetic prosthetic hand: Design, manufacturing and preliminary examination*. Paper presented at the 2018 IEEE/RSJ international conference on intelligent robots and systems (IROS).
2. Dollar, A. M., & Howe, R. D. (2007, June 13–15). *The SDM hand as a prosthetic terminal device: A feasibility study*. Paper presented at the 2007 IEEE 10th international conference on rehabilitation robotics.
3. Takeda, H., Tsujiuchi, N., Koizumi, T., Kan, H., Hirano, M., & Nakamura, Y. (2009, September 3–6). *Development of prosthetic arm with pneumatic prosthetic hand and tendon-driven wrist*. Paper presented at the 2009 annual international conference of the IEEE engineering in medicine and biology society.

4. Dalley, S. A., Wiste, T. E., Varol, H. A., & Goldfarb, M. (2010, August 31–September 4). *A multigrasp hand prosthesis for transradial amputees*. Paper presented at the 2010 annual international conference of the IEEE engineering in medicine and biology.
5. Nagase, J.-Y., Wakimoto, S., Satoh, T., Saga, N., & Suzumori, K. (2011). Design of a variable-stiffness robotic hand using pneumatic soft rubber actuators. *Smart Materials and Structures, 20*(10), 105015. https://doi.org/10.1088/0964-1726/20/10/105015
6. Tondu, B. (2012). Modelling of the McKibben artificial muscle: A review. *Journal of Intelligent Material Systems and Structures, 23*(3), 225–253. https://doi.org/10.1177/1045389X11435435
7. Sun, Y., Song, Y. S., & Paik, J. (2013, November 3–7). *Characterization of silicone rubber based soft pneumatic actuators*. Paper presented at the 2013 IEEE/RSJ international conference on intelligent robots and systems.
8. Galloway, K. C., Polygerinos, P., Walsh, C. J., & Wood, R. J. (2013, November 25–29). *Mechanically programmable bend radius for fiber-reinforced soft actuators*. Paper presented at the 2013 16th international conference on advanced robotics (ICAR).
9. Maeder-York, P., Clites, T., Boggs, E., Neff, R., Polygerinos, P., Holland, D., … Walsh, C. J. (2014, April 8–10). *Biologically inspired soft robot for thumb rehabilitation*. Paper presented at the ASME design of medical devices conference, Minneapolis, MN.
10. O'Neill, C. (2014). *An Advanced, low cost prosthetic arm*. Paper presented at the IEEE SENSORS 2014 proceedings.
11. Mosadegh, B., Polygerinos, P., Keplinger, C., Wennstedt, S., Shepherd, R. F., Gupta, U., … Whitesides, G. M. (2014). Pneumatic networks for soft robotics that actuate rapidly. *Advanced Functional Materials, 24*(15), 2163–2170. https://doi.org/10.1002/adfm.201303288
12. Connolly, F., Polygerinos, P., Walsh, C. J., & Bertoldi, K. (2015). Mechanical programming of soft actuators by varying fiber angle. *Soft Robotics, 2*(1), 26–32. https://doi.org/10.1089/soro.2015.0001
13. Stopforth, R., Riet, D., Bright, G., & Diegel, O. (2015). A low cost design of a 3D printed multi-fingered myoelectric prosthetic hand. In E. Brusa (Ed.), *Mechatronics: Principles, technologies and applications* (pp. 85–117). Nova Science Publishers, Inc.
14. Gretsch, K. F., Lather, H. D., Peddada, K. V., Deeken, C. R., Wall, L. B., & Goldfarb, C. A. (2016). Development of novel 3D-printed robotic prosthetic for transradial amputees. *Prosthetics and Orthotics International, 40*(3), 400–403. https://doi.org/10.1177/0309364615579317
15. Beyrouthy, T., Kork, S. K. A., Korbane, J. A., & Abdulmonem, A. (2016, August 3–6). *EEG mind controlled smart prosthetic arm*. Paper presented at the 2016 IEEE international conference on emerging technologies and innovative business practices for the transformation of societies (EmergiTech).
16. Cañizares, A., Pazos, J., & Benítez, D. (2017, November 8–10). *On the use of 3D printing technology towards the development of a low-cost robotic prosthetic arm*. Paper presented at the 2017 IEEE international autumn meeting on power, electronics and computing (ROPEC).
17. Zuniga, J. M., Carson, A. M., Peck, J. M., Kalina, T., Srivastava, R. M., & Peck, K. (2017). The development of a low-cost three-dimensional printed shoulder, arm, and hand prostheses for children. *Prosthetics and Orthotics International, 41*(2), 205–209. https://doi.org/10.1177/0309364616640947
18. Connolly, F., Walsh, C. J., & Bertoldi, K. (2017). Automatic design of fiber-reinforced soft actuators for trajectory matching. *Proceedings of the National Academy of Sciences, 114*(1), 51–56. https://doi.org/10.1073/pnas.1615140114
19. Mai, N.-A., Duc Tang, T., Trung, V., Le, C. H., Tu, N., Minh, H., & Quoc, L. (2018). *Cost-effective design and development of a prosthetic hand* (Vol. 69, pp. 225–229). Springer.
20. Mata Amritanandamayi, D., Udupa, G., & Sreedharan, P. (2018). A novel underactuated multi-fingered soft robotic hand for prosthetic application. *Robotics and Autonomous Systems, 100*, 267–277. https://doi.org/10.1016/j.robot.2017.11.005
21. Mick, S., Lapeyre, M., Rouanet, P., Halgand, C., Benois-Pineau, J., Paclet, F., … de Rugy, A. (2019). Reachy, a 3D-printed human-like robotic arm as a testbed for human-robot control. *Strategies, 13*(65). https://doi.org/10.3389/fnbot.2019.00065

22. Sureshbabu, A. V., Rass, D., & Zimmermann, M. (2019, December 2–6). *A lightweight transradial hand prosthesis with a variable position thumb and thermoregulation*. Paper presented at the 2019 19th international conference on advanced robotics (ICAR).
23. Barrie, D. D., Margetts, R., & Goher, K. (2020). SIMPA: Soft-grasp infant myoelectric prosthetic arm. *IEEE Robotics and Automation Letters, 5*(2), 699–704. https://doi.org/10.1109/LRA.2019.2963820
24. McDonald, M., Petropoulakis, L., Soraghan, J., Lavery, L., Waddell, J., Walker, K., & Macdonald, T. (2020). Design and control of a low cost myoelectric prosthetic hand. *International Journal of Mechatronics and Automation, 7*, 94. https://doi.org/10.1504/IJMA.2020.10030739
25. Fras, J., Glowka, J., & Althoefer, K. (2020, May 15–July 15). *Instant soft robot: A simple recipe for quick and easy manufacturing*. Paper presented at the 2020 3rd IEEE international conference on soft robotics (RoboSoft).
26. ISHN. (2014). *Statistics on hand and arm loss*. Publication no. https://www.ishn.com/articles/97844-statistics-on-hand-and-arm-loss#:~:text=transradial%20amputations%20make%20up%2060%25%20of%20total%20wrist%20and%20hand%20amputations&text=In%20US%2041%2C000%20persons%20are,under%2021%20years%20of%20age). from Industrial Safety and Hygiene News.
27. Verma, D., Dash, P., Bhaskar, S., Pal, R. P., Jain, K., Srivastava, R. P., et al. (2016). *Disabled persons in India: A statistical profile 2016* (pp. 1–107). Ministry of Statistics and Programme Implementation, Government of India.

Macro-Mechanical Modeling of 3D Printed Material

Ashutosh Mishra, Abhishek Kumar Tiwari, Rakesh Kumar, and Wasim Ashraf

1 Introduction

Historically, manufacturing was one of the important activities' humans were involved into, since the beginning of human civilization, where stone hunting tools, mud containers, etc., were produced [1]. Manufacturing as a broad term involves the conversion of raw material into useful product with desired functionality. A significant portion of a nations' gross domestic product (GDP) depends weweon the contributions from manufacturing industries. Developing countries, e.g., India, are laying greater interest in manufacturing sector in recent years. With the rapid advancement in technologies, new manufacturing processes are developed focusing on various aspects such as reduction in manufacturing time, use of hard to hardest material for product development, reduction of material loss, etc. In this regard, a greater emphasis is being given to the research and development activities to solve the issues associated with the existing technologies to manufacture products with enhanced properties and reduced manufacturing time [2]. Increasing global competition in the manufacturing sector has led to the rapid development efforts in the field of advance manufacturing processes such as additive manufacturing (AM), which may reduce the overall manufacturing cost and time [3]. There is a need to understand the behavior of the products manufactured through the process of additive manufacturing. Therein, the current chapter describes an additive

A. Mishra · A. K. Tiwari (✉) · W. Ashraf
Department of Applied Mechanics, Motilal Nehru National Institute of Technology Allahabad, Allahabad, Uttar Pradesh, India
e-mail: amishra@mnnit.ac.in; aktiwari@mnnit.ac.in

R. Kumar
2Department of Mechanical Engineering, Manipal University Jaipur, Jaipur, Rajasthan, India
e-mail: kumar.rakesh@jaipur.manipal.edu

C. Prakash et al. (eds.), *Additive, Subtractive, and Hybrid Technologies*,
Mechanical Engineering Series, https://doi.org/10.1007/978-3-030-99569-0_11

manufacturing process and its various governing parameters which define the quality of the additively manufactured product. The chapter mainly aimed to cover the aspects of numerical modeling and analysis of the additively manufactured product.

Conventional manufacturing processes of material removal for manufacturing of products are still very popular and are still used in modern days for variety of products [4]. With the advent of computers and computer-aided design (CAD) technologies, it has been observed that non-conventional manufacturing processes are gaining wide popularity among the manufacturing industries. As one of the most advanced manufacturing technologies, AM has emerged into the current manufacturing dominant world with several beneficial features [5]. It is of significant interest to common people to know how AM can be used and what are the different kind of new products that can be developed using this process. In addition, it is gaining attention among researchers worldwide with several other names such as rapid prototyping, three-dimensional (3D) printing, layered manufacturing, generative manufacturing. Out of all the names of the AM, 3D printing is a most common name and well-known term. The underlying principle of 3D printing is to build a 3D actual object from several layers of uniform thickness (mostly) [6]. The principle of the 3D printing and other working procedure is detailed in the next section.

2 Three-Dimensional (3D) Printing

Three-dimensional (3D) printing process belongs to the additive or generative manufacturing processes unlike material removal or forming processes such as lath machining, grinding, milling or coining, etc. in which the desired form is shaped by material removal or plastic deformation. The layers are contoured as per the recorded or desired 3D data sets and then placed layer by layer over one another to prepare the final part. The 3D data sets are typically obtained by 3D CAD process or by utilizing some scanning technologies, for example, computerized tomography scanning technique (CT scanning). 3D printing is broadly categorized into three classes as shown in Fig. 1.

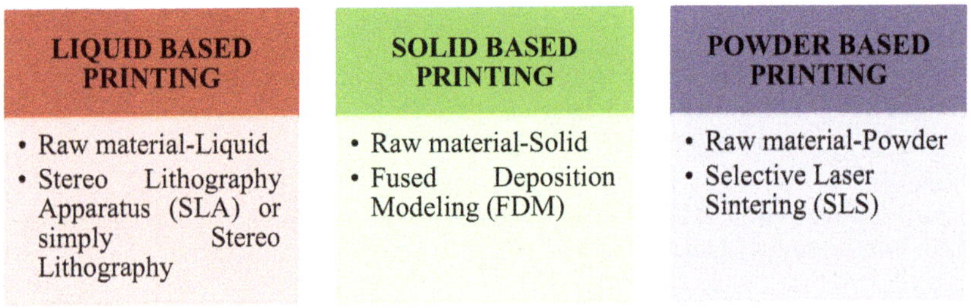

Fig. 1 Classification of 3D printing

3 Fused Deposition Modeling

Liquid-based printing such as stereo lithography and powder-based printing such as selective laser sintering (SLS) use liquid and powder raw materials, respectively. The chapter covers the macro-mechanical modeling procedure of parts developed with fused deposition modeling (FDM) as a 3D printing process. FDM is one of the 3D printing technologies recognized widely owing to its simplicity and minimum time of realization of a new product. FDM as a solid-based 3D printing technology uses a solid wire of polymer, metal-infused polymer, or pure metals. Plastic is a popular choice as FDM material for its quick and easy deposition possibilities to obtain the finished product. For low-volume manufacturing of complex end use and rapid prototype development, demand of FDM is increasing rapidly. Some FDM printers use metal as a fused material, but finished product may not be fully dense compared to that obtained by using casting process. FDM metal products are not commercialized in large scale as of now, but popularity of FDM can be attributed to its advantages over other processes [7].

Advantages of Fused Deposition Modeling

- Many types of material can be used.
- Support structure is not required.
- Economical setup.
- No post-curing is required.

Disadvantages of Fused Deposition Modeling

- Porosity and rough surface finish are obtained.
- Product may become weak against loading in Z-direction.
- Very difficult to remove support parts.
- Takes longer time to fabricate large parts.

3.1 Principle and Working of Fused Deposition Modeling

In FDM, a filament is unwound and feed material to the extrusion nozzle provided with heater to melt the filament at melting temperature. The nozzle provides the material deposition in *x-y plane* tracing the contour of a layer on a table which moves in *z*-direction. Once a layer is deposited so as to get another layer deposited over it [8]. The FDM setup is capable to move in *x, y,* and *z* axes, depositing one layer at a time and stacking other layers over as shown in Fig. 2a, b. The process of extrusion of fused material deposition is very similar to the way a melted bead of glue is produced from a hot glue gun nozzle. The result of the solidified material piling layer by layer is a 3D model which is then surface finished as per the requirement by using appropriate surface finishing method.

Production of the parts with unique characteristics is possible through the FDM process [9] and directional deposition of material though the FDM nozzle imparts

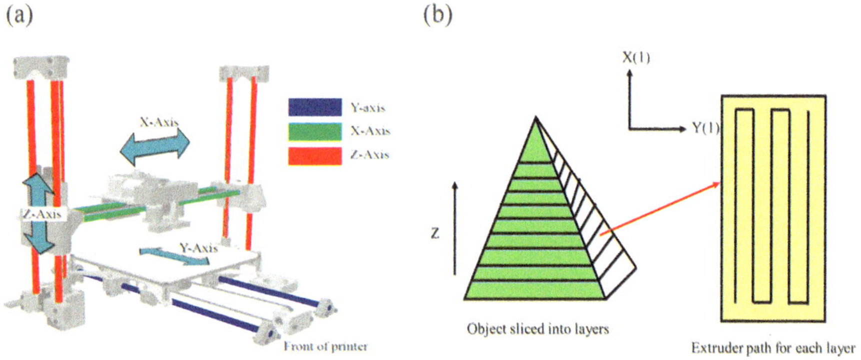

Fig. 2 (**a**) Schematic 3D printing setup and (**b**) 3D object slicing and path generation

anisotropic properties in the parts produced. The utmost importance of 3D printing is the adherence of successive layers which is achieved by the chemical bond due to the solidification of molten layers by the application of some adhesives. The layer-by-layer deposition of fused material which yield to a final product is the reason for assuming an FDM part as a composite material, whereby the filament material deposition defines the strength and quality of the final part. It is therefore worthwhile to discuss various process parameters which govern the functionality and durability of FDM parts/products and present the numerical procedure to predict the behavior of the parts.

3.2 Governing Parameters

FDM technology is highly flexible, and its integration with most of the available CAD software is very easy. As the first step, 3D solid model of a desired part is generated using any available CAD software which is then provided to the FDM machine's software in stereo lithography (STL) format. The STL format is compatible with most of the modern available CAD software which discretizes the part into smaller tessellations and thus simplifies the geometry of the part. The tessellation strategies and geometry of deposition path of a layer has a significant effect on the resulting mechanical behavior of the 3D printed parts [10]. In addition, the mechanical properties of FDM parts also get quantitatively affected by some governing parameters listed below:

1. *Bead width*: This denotes the thickness of the bead of molten material that the nozzle of FDM machine deposits. The width may vary for different nozzles used in FDM [11]. The distance between the beads deposited (or road gaps) also governs the properties of produced parts as decreasing the road gaps lead to an increase in the layer adhesion [11].

2. *Air gap*: It is the space between the beads which is assumed to be zero, such that the adjacent beads touch each other. This space can be varied to make the gap, positive or negative also, which means that the beads of material do not touch, or they partially occupy the same space [12]. Positive air gap results in a loosely packed structure which takes less time to prepare while results in a dense structure, which requires a longer build time.
3. *Temperature*: It is the temperature of the fused filament of the model material which is used to control the flow and deposition speed with the movement of nozzle.
4. *Raster orientation*: It is the deposition direction of the fused filament material (roads).

Fused deposited filament roads bears a cross section varying from flat to round appearance which is governed by the nozzle geometry, flow rate of the fused material, and temperature (Fig. 3). Thus, the perfectly rectangular filament deposition is almost impossible to be achieved. Due to such imperfect geometry of cross section of the deposited material, the sharp corners are not achieved, and thus, the voids are created between each roads of the filament [13]. These voids are generated with each layer-to-layer deposition and are responsible for strength reduction due to the increasing air gap. Apart from the parameters discussed above, the material behavior of 3D printed parts is largely dependent on the choice of the materials used for printing the part and the process adopted for printing [14]. The present chapter however only discusses the modeling procedure of FDM parts made up of PLA materials, though the same understanding can be extended for modeling with other materials [15].

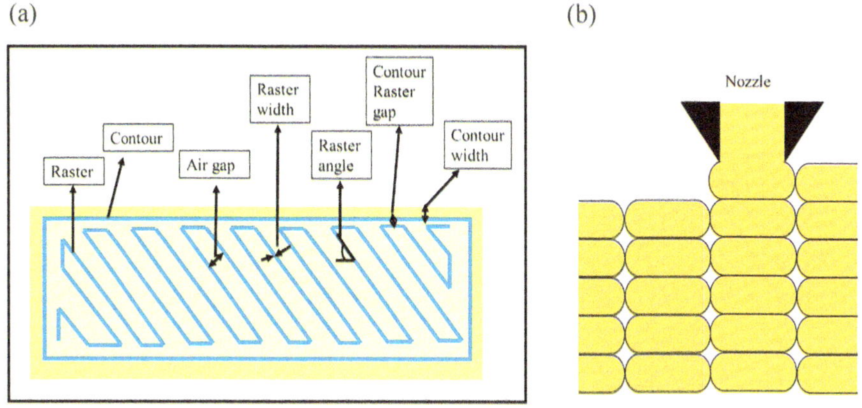

Fig. 3 (**a**) FDM build parameters; (**b**) filament bead/road with air gaps

4 Constitutive Modeling

The numerical modeling of deformation behavior of FDM parts involves multifaceted concepts depending upon the complexity and accuracy considerations. Material modeling of FDM parts is widely popular in the current age of modern computation facilities, and it needs some fundamental understanding of macro-mechanical modeling of composites.

The complex procedure of materials behavior prediction of products prepared by FDM is due to the effect of process parameters such as material deposition strategy, deposited path, layer thickness, etc. These processes are responsible for the anisotropic properties in the resulting FDM products. It is demonstrated in the literature that the mechanical properties of the filament material in the FDM process [16] are different from the finished FDM products additively manufactured by using the same filament. It can be seen that the researchers initially started out with improving and optimizing the deposition process (contoured or raster deposition method as shown in Fig. 4. This was done to increase the product quality in terms of both strength and aesthetics.

Ahn et al. [9] identified the effect of various governing parameters of FDM parts and demonstrated the significance of design strategy and process variables on the strength. Compression tests were also performed to understand the effect of build direction which showed higher compressive strength compared to the tensile strengths. Authors used test specimens produced by injection molding and compare the material properties of it with the 12-layered printed specimens for the different loading cases. Layer orientation was both unidirectional and bidirectional for 12 layers for each load cases. A detailed review of significance of process parameters in FDM parts is presented by Mohan et al. [17]. Various works done to predict the behavior of 3D printed parts demonstrated that composite lamina approach can suitably be applied to simulate the behavior of FDM printed parts. Present work describes the two approaches to predict the deformation behavior of 3D printed FDM parts.

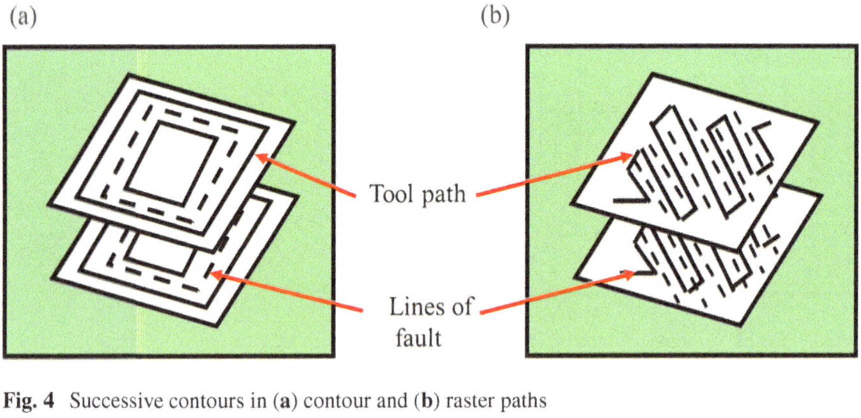

Fig. 4 Successive contours in (**a**) contour and (**b**) raster paths

4.1 Orthotropic Composite Lamina Approach

A single layer deposited for 3D printing to produce a part using FDM can be assumed as a lamina. The mechanical behavior of 3D printed layers can be understood using lamina approach. A lamina can be flat or curved and may have unidirectional or multidirectional fiber orientation. In order to predict the material behavior of 3D printed parts, several approaches are suggested. For example, Bellini and Guceri [18] used orthotropic material modeling approach as defined below:

$$\{\varepsilon\} = [S]\{\sigma\} \tag{1}$$

where *S* is a compliance matrix so the constitutive relation for orthotropic material can be written as below:

$$\begin{Bmatrix} \varepsilon_{11} \\ \varepsilon_{22} \\ \varepsilon_{33} \\ \gamma_{23} \\ \gamma_{13} \\ \gamma_{12} \end{Bmatrix} = \begin{bmatrix} S_{11} & S_{12} & S_{13} & 0 & 0 & 0 \\ S_{12} & S_{22} & S_{23} & 0 & 0 & 0 \\ S_{13} & S_{23} & S_{33} & 0 & 0 & 0 \\ 0 & 0 & 0 & S_{44} & 0 & 0 \\ 0 & 0 & 0 & 0 & S_{55} & 0 \\ 0 & 0 & 0 & 0 & 0 & S_{66} \end{bmatrix} \begin{Bmatrix} \sigma_{11} \\ \sigma_{22} \\ \sigma_{33} \\ \tau_{23} \\ \tau_{13} \\ \tau_{12} \end{Bmatrix} \tag{2}$$

$$\begin{Bmatrix} \varepsilon_{11} \\ \varepsilon_{22} \\ \varepsilon_{33} \\ \gamma_{23} \\ \gamma_{31} \\ \gamma_{12} \end{Bmatrix} = \begin{bmatrix} \frac{1}{E_1} & -\frac{\nu_{12}}{E_1} & -\frac{\nu_{13}}{E_1} & 0 & 0 & 0 \\ -\frac{\nu_{12}}{E_1} & \frac{1}{E_2} & -\frac{\nu_{23}}{E_2} & 0 & 0 & 0 \\ -\frac{\nu_{13}}{E_1} & -\frac{\nu_{23}}{E_2} & \frac{1}{E_3} & 0 & 0 & 0 \\ 0 & 0 & 0 & \frac{1}{G_{23}} & 0 & 0 \\ 0 & 0 & 0 & 0 & \frac{1}{G_{31}} & 0 \\ 0 & 0 & 0 & 0 & 0 & \frac{1}{G_{12}} \end{bmatrix} \begin{Bmatrix} \sigma_{11} \\ \sigma_{22} \\ \sigma_{33} \\ \tau_{23} \\ \tau_{31} \\ \tau_{12} \end{Bmatrix} \tag{3}$$

where, ε_{11}, ε_{22}, ε_{33} are normal strains, γ_{23}, γ_{31}, γ_{12} are shear strains, σ_{11}, σ_{22}, σ_{33} are normal stresses, τ_{23}, τ_{31}, τ_{12} are shear stresses E_1, E_2, E_3 are Young's moduli, G_{23},G_{31},G_{12}, are shear moduli, and, ν_{12}, ν_{23}, ν_{13} are Poisson's ratio.

It is to be noted that 1 and 2 are subscripts for a unidirectional lamina and are the directions along the fibers and across the fibers, respectively. It is necessary to obtain nine independent constants for orthotropic material modeling, while Young's moduli and Poisson's ratios are related as:

$$\frac{\nu_{21}}{E_2}=\frac{\nu_{12}}{E_1},\frac{\nu_{31}}{E_3}=\frac{\nu_{13}}{E_1},\frac{\nu_{23}}{E_2}=\frac{\nu_{32}}{E_3}$$

or one can write this relation in index form as:

$$\frac{\nu_{ij}}{E_i}=\frac{\nu_{ji}}{E_j}\ (\text{nosumover}\, i \text{ and } j)$$

and

$$G_{23}=\frac{E_2}{2(1+\nu_{23})},G_{31}=\frac{E_3}{2(1+\nu_{31})},G_{12}=\frac{E_1}{2(1+\nu_{12})}$$

or in general as:

$$G_{ij}=\frac{E_x}{2(1+\nu_{xy})} \tag{4}$$

where x is the direction of the load applied, y is the perpendicular direction, and i–j corresponds to 1–2 for lamina built in xy plane, 3–1 for lamina in zx plane, and 2–3 for lamina in yz plane. The compliance matrix shown above is symmetric as the stiffness matrix for orthotropic materials is also symmetric. The stiffness matrix for the present case is obtained by inverting the compliance matrix as:

$$\begin{Bmatrix}\sigma_{11}\\ \sigma_{22}\\ \sigma_{33}\\ \tau_{23}\\ \tau_{13}\\ \tau_{12}\end{Bmatrix}=\begin{bmatrix}\frac{1-\nu_{23}\nu_{32}}{E_2E_3\Delta} & \frac{\nu_{21}+\nu_{23}\nu_{31}}{E_2E_3\Delta} & \frac{\nu_{31}+\nu_{21}\nu_{32}}{E_2E_3\Delta} & 0 & 0 & 0\\ & \frac{1-\nu_{13}\nu_{31}}{E_1E_3\Delta} & \frac{\nu_{32}+\nu_{12}\nu_{31}}{E_1E_3\Delta} & 0 & 0 & 0\\ & & \frac{1-\nu_{12}\nu_{21}}{E_1E_2\Delta} & 0 & 0 & 0\\ & & & G_{23} & 0 & 0\\ & \text{symmetric} & & & G_{13} & 0\\ & & & & & G_{12}\end{bmatrix}\begin{Bmatrix}\varepsilon_{11}\\ \varepsilon_{22}\\ \varepsilon_{33}\\ \gamma_{23}\\ \gamma_{13}\\ \gamma_{12}\end{Bmatrix} \tag{5}$$

where

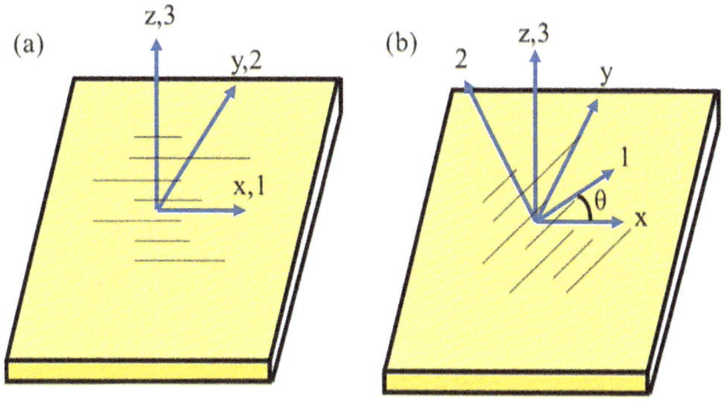

Fig. 5 Coordinate system in (**a**) specially orthotropic model and (**b**) generally orthotropic model

$$\Delta = \frac{\left(1 - \nu_{12}\nu_{21} - \nu_{23}\nu_{32} - \nu_{13}\nu_{31} - 2\nu_{21}\nu_{32}\nu_{13}\right)}{\left(E_1 E_2 E_3\right)}$$

The orientation of a single lamina can be different with respect to laminate (which is defined relative to the global coordinate system). Therefore, mathematical modeling of the FDM parts can be done by adopting two model systems defined as: (1) specially orthotropic model and (2) generally orthotropic model which has been represented in Fig. 5. Specially, orthotropic model is the one with global coordinate system X-Y-Z aligned ($\theta = 0°$) with the local material coordinate system 1-2-3 as defined by Bellini and Guceri [18], which is otherwise also called as generally orthotropic material model. A single layer of FDM part can be considered as a composite lamina with especially orthotropic or generally orthotropic layer depending upon the raster angle of the deposited fused material. The single layer when combined together to form a part can be assumed as laminates; therefore, the theory of laminates becomes applicable for simulating the behavior of the FDM parts.

4.2 *Transversely Isotropic Composite Lamina Approach*

A transversely isotropic material is one with physical properties that are symmetric about an axis that is normal to a plane of isotropy. This transverse plane has infinite planes of symmetry, and thus, within this plane, the material properties are the same in all directions. Therefore, such materials are also known as "polar anisotropic" materials. Unidirectional lamina is typically considered to behave as transversely isotropic material [19]. The elastic moduli for a transversely isotropic lamina are same for lateral and transverse directions. The stiffness matrix for transversely isotropic material is shown below, and the corresponding compliance matrix is given defined in Eqs. (6) and (7) in such a way that the plane of isotropy is along plane 1–2.

$$[C_{ij}] = \begin{bmatrix} C_{11} & C_{12} & C_{13} & 0 & 0 & 0 \\ C_{12} & C_{11} & C_{13} & 0 & 0 & 0 \\ C_{13} & C_{13} & C_{33} & 0 & 0 & 0 \\ 0 & 0 & 0 & C_{44} & 0 & 0 \\ 0 & 0 & 0 & 0 & C_{44} & 0 \\ 0 & 0 & 0 & 0 & 0 & C_{66} \end{bmatrix} \quad (6)$$

$$[S_{ij}] = \begin{bmatrix} S_{11} & S_{12} & S_{13} & 0 & 0 & 0 \\ S_{12} & S_{11} & S_{13} & 0 & 0 & 0 \\ S_{13} & S_{13} & S_{33} & 0 & 0 & 0 \\ 0 & 0 & 0 & S_{44} & 0 & 0 \\ 0 & 0 & 0 & 0 & S_{44} & 0 \\ 0 & 0 & 0 & 0 & 0 & 2(S_{11} - S_{12}) \end{bmatrix} = [C_{ij}]^{-1} \quad (7)$$

Various elements of the compliance matrix can be obtained by equating $E_{11} = E_{22}$, $\nu_{13} = \nu_{12}$ and $G_{31} = G_{12}$ in Eq. (5). The constitutive relation for unidirectional lamina therefore requires the values of five material properties and thus further simplifies the numerical modeling of FDM parts [20].

5 Macro-Mechanical Modeling and Analysis of FDM Parts

FDM parts are constructed with stacking several layers in the direction of thickness. Therefore, the composite laminate approach is adopted to perform the macro-mechanical modeling and analysis of the FDM printed specimen.

5.1 Specimen Fabrication

The tensile test specimens are fabricated according to ASTM standard D638 Type IV using a CREALITY CR-10S model FDM machine. The schematic diagram of the FDM specimen is shown in Fig. 6a. All dimensions shown in Fig. 6a are according to the ASTM standard. The length, width, and thickness of the specimen are 115 mm, 19 mm, and 4 mm, respectively. The specimens are made by using Polylactic Acid (PLA) material which is environment friendly and easily available with low cost. The specimens are prepared with the three different deposition strategy as shown in Fig. 6.

The three different deposition strategies for the specimen are achieved by considering the three deposition angles with a fixed nozzle diameter of 0.4 mm and other

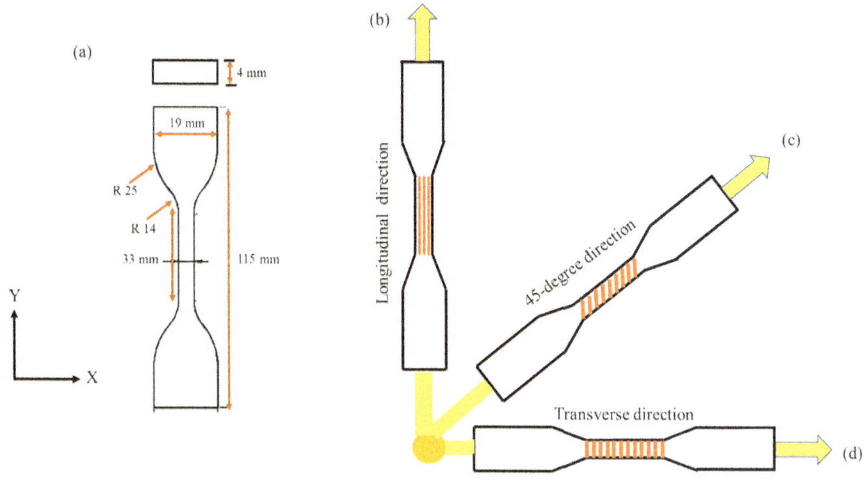

Fig. 6 (**a**) The schematic diagram for the tensile test specimen, (**b**) longitudinal deposition strategy, (**c**) 45 degree deposition strategy, (**d**) transverse deposition strategy

Table 1 Process parameter for 3D printed part

Parameters	Value
Deposition angles	0°, 45°, and 90°
Nozzle diameter	0.4 mm
Layer thickness	0.3 mm
Infill percentage	100%
Printing temperature	205 °C
Bed temperature	60 °C
Printing speed	45 mm/s
Outline contour	2
Material	PLA
Contouring type	Unidirectional

parameters as shown in the Table 1, though specimen shown in Fig. 7 are for two deposition angles only.

5.2 *Mathematical Modeling*

Figure 7 shows two specimens with 45° and 0° deposition angle to highlight the beads and contours (here x axis is in the transverse direction, i.e., along 1 while y-axis is in the longitudinal direction, i.e., along 2) considered for the current macro-mechanical analysis work. The effect of initial contouring and the small length of

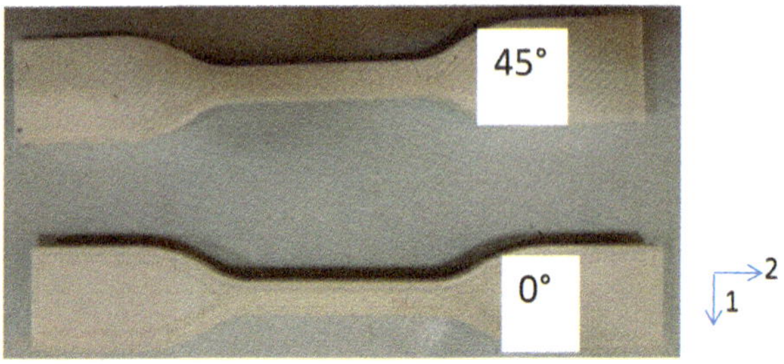

Fig. 7 FDM specimen fabricated at 0° and 45°

beads deposited while changing the direction of nozzle for continuous deposition is neglected in the present analysis. The framework of the present analysis of FDM part is based on the composite laminate approach. In this approach, plane stress approximations are suitably applied in which results into the following:

$$\sigma_{33} = \sigma_{23} = \sigma_{13} = 0, \gamma_{31} = \gamma_{23} = 0 \quad \text{and} \quad \gamma_{21} = \frac{\tau_{12}}{G_{12}}$$

Analytical solutions of the loaded FDM printed specimen are obtained by using classical laminate theory (CLT) and modified classical laminate theory (MCLT) approach which is described in the next sub-section.

5.3 *Classical Laminate Theory (CLT)*

CLT assumes a perfect bonding between different layers of the laminate and also that there is no void or air gap present in the parts. This means 100% filling of the material is achieved during FDM. Perfect bonding means that there are no flaws or gaps between the various layers. Additionally, there is no relative slip between the adjacent lamina. This theory is only valid for thin layers and small displacement in the transverse direction of the FDM parts. The important constitutive relations of CLT applied for each layer in the present analysis can be expressed as Eqs. (9) and (10) assuming transversely isotropic composite approach.

$$\begin{Bmatrix} \varepsilon_{11} \\ \varepsilon_{22} \\ \gamma_{12} \end{Bmatrix} = \begin{vmatrix} S_{11} & S_{12} & 0 \\ S_{12} & S_{22} & 0 \\ 0 & 0 & S_{66} \end{vmatrix} \begin{Bmatrix} \sigma_{11} \\ \sigma_{22} \\ \tau_{12} \end{Bmatrix} \tag{8}$$

$$\begin{Bmatrix} \sigma_{11} \\ \sigma_{22} \\ \tau_{12} \end{Bmatrix} = [Q] \begin{Bmatrix} \varepsilon_{11} \\ \varepsilon_{22} \\ \varepsilon_{12} \end{Bmatrix} \tag{9}$$

$$\begin{Bmatrix} \sigma_{xx} \\ \sigma_{yy} \\ \sigma_{xy} \end{Bmatrix}_k = \left[\bar{Q}\right] \begin{Bmatrix} \varepsilon_{xx} \\ \varepsilon_{yy} \\ \gamma_{xy} \end{Bmatrix}_k \quad (10)$$

where $[Q]$ and $\left[\bar{Q}\right]$ are the reduced stiffness matrix and transformed reduced stiffness matrix, respectively.

The stress and strain values with reference to the global coordinate system for FDM specimen are obtained using Eqs. (8)–(10). It is to be noted that the number of independent material constants are reduced to four, i.e., E_1, E_2, ν_{12}, G_{12}. The elements of matrix $[Q]$ can be obtained using Eq. (5) while elements of $\left[\bar{Q}\right]$, i.e., kth layer of the FDM part, is obtained by using Eq. (11) and the transformation matrix $[T]$ [21]. The stress for each layer can be added/integrated over an area to predict the total stress. CLT approach assumes 100% fill, and therefore, a single layer is considered to analyze the unidirectional FDM part in the present analysis.

$$\left[\bar{Q}\right] = [T]^{-1}[Q][T]^{-T} \quad (11)$$

5.4 *Modified Classical Laminate Theory (MCLT)*

CLT assumes perfect bonding between layers, i.e., no void and gaps are present between fibers or layers which is not realistic in case of FDM parts as some void and gap are always present. The presence of voids and air gaps has prominent effect in the property of the FDM part so modified set of equations are defined to determine the material properties to account for the effect of gaps and voids. The effect of the voids and air gaps is visible and shown in the results section when MCLT is used to analyze the behavior of the FDM parts.

According to MCLT, the material properties are expressed as below:

$$E_{11} = (1 - \rho_1)E_{\text{pl}} \quad (12)$$

$$E_{22} = \varphi(1 - \rho_2)E_{\text{pl}} \quad (13)$$

$$G_{12} = \varphi'(1 - \rho_2)G \quad (14)$$

where ρ_1 is area void density, and ρ_2 is linear void ratio, whereas φ and φ' are empirical factors that takes into account the bonding strength which varies from 0 to 1, E_{pl} is elastic modulus of plastic filament, G_{12} is shear modulus value of the laminate, G is shear modulus of plastic material, E_{11} is elastic modulus in direction 1 in local coordinate system, and E_{22} is elastic modulus in direction 2 in the local coordinate system. Void analysis is performed to estimate the variables ρ_1 and ρ_2. This requires

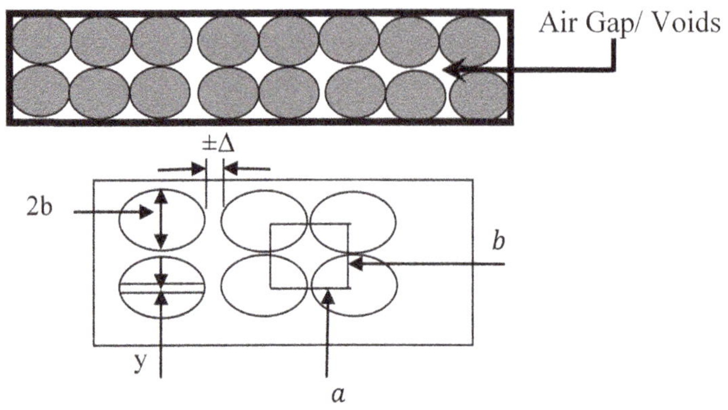

Fig. 8 Schematic representation of gaps and measurement, i.e., *a*, *b*, and *Δ*

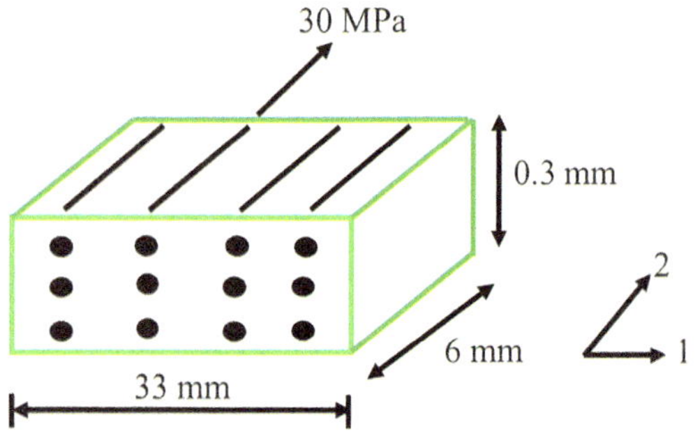

Fig. 9 Geometry of the specimen for analytical solution using CLT and MCLT

the slicing of FDM parts and a careful examination on the gaps between fibers and beads. In addition, a mathematical relation is also used to calculate the parameters which is defined as:

$$\rho_1 = 1 - \frac{\pi ab}{(2b-\delta)(2a \pm \Delta)} \text{and} \rho_2 = \frac{b-y}{b} \tag{15}$$

Area void density is in the plane normal to the filament or strands and linear void fraction along the transverse direction. *a*, *Δ*, and *y* used in Eq. (15) are represented in Fig. 8.

5.5 Analytical Modeling for Monotonic Loading

Analytical modeling is performed by considering an FDM specimen of dimensions shown in Fig. 9, and a loading is done with a 30 MPa normal stress along fiber orientation of 90°.

5.6 Constants

Considering transversely isotropic plane stress condition with plane 2–3 as a plane of symmetry the material constants are listed in Table 2 for CLT and MCLT [22].

5.7 Finite Element Analysis

The finite element analysis (FEA) is conducted in ABAQUS Software [23] using monotonic load for three deposition strategy at 0°, 45°, and 90°. The results are compared with analytical results. It is to be noted that the FEA performed for CLT and MCLT uses the same loading conditions for the part model tested. Once the part modeling is done in ABAQUS, the material constants listed in Table 2 for CLT and MCLT are supplied as input in the material definition module. The S4R-type element with four nodes is used to prepare the meshed model with total number of 1742 elements. The ply stack up of the composite laminates are shown in Fig. 10, representing the FDM specimen of 14 layers is subjected to 30 MPa of stress load.

Table 2 Constants used in laminate theory analysis

Parameters	Value	
	CLT	MCLT
E_{11}	40 GPa	38.8 GPa
$E_{22} = E_{33}$	8 GPa	5.40672 GPa
$G_{12} = G_{13}$	4 GPa	2.3936 GPa
G_{23}	3.077 GPa	3.077 GPa
$\nu_{12} = \nu_{13}$	0.25	0.25
ν_{23}	0.3	0.3
ρ_1	–	0.0346
ρ_2	–	0.2776
φ	–	0.96
φ'	–	0.85

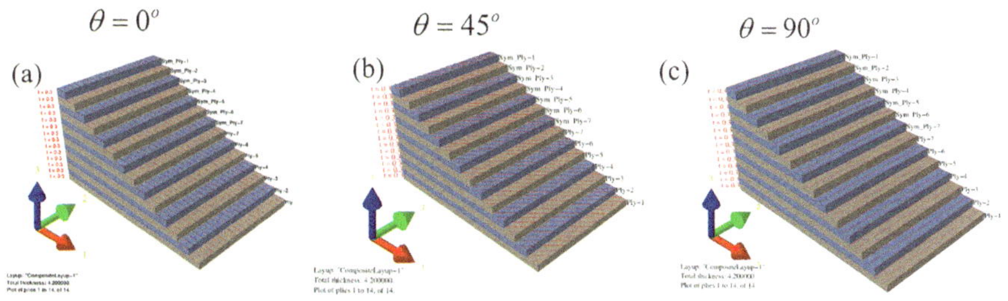

Fig. 10 Ply stack up for (**a**) 0°, (**b**) 45°, and (**c**) 90° deposition angle for both CLT and MCLT approaches

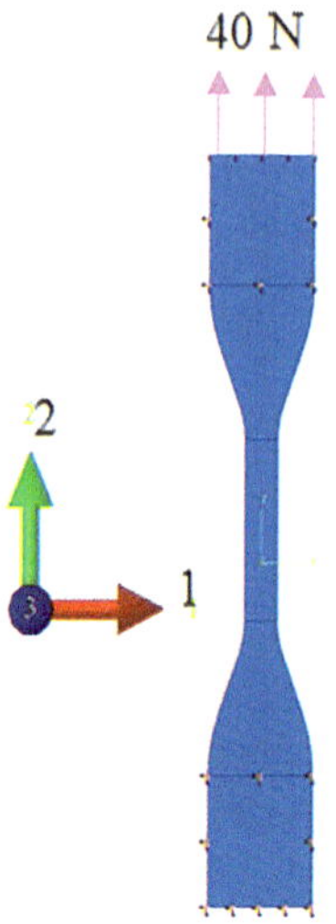

Fig. 11 Specimen load and boundary conditions

This stress load is generated due to the concentrated 40 N load with free end unconstrained only in the direction of load.

Tensile loading for all the three deposition strategy is applied on the specimen with the geometry having same gauge length and other dimensions as considered in analytical method. The specimen is fixed at one end while a load of 40 N (maximum stress generated 30 MPa in gauge length) is applied at the other end as shown in Fig. 11.

6 Results and Discussion

The macro-mechanical modeling is performed to obtain the analytical and numerical results while considering the two theories, i.e., CLT and MCLT.

6.1 Analytical Results of CLT

The relationship used to compute the strain is defined as:

$$\begin{Bmatrix} \varepsilon_{11} \\ \varepsilon_{22} \\ \gamma_{12} \end{Bmatrix} = \begin{bmatrix} \frac{1}{40} & \frac{-0.25}{40} & 0 \\ \frac{-0.25}{40} & \frac{1}{8} & 0 \\ 0 & 0 & \frac{1}{4} \end{bmatrix} \begin{Bmatrix} 30 \\ 0 \\ 0 \end{Bmatrix} 10^{-3} = \begin{Bmatrix} 0.00075 \\ -0.0001875 \\ 0 \end{Bmatrix} \tag{16}$$

$$\text{or}\, \varepsilon_{22} = -0.0001875$$

Similarly, strains ε_{22} are obtained for 0° and 45° fiber orientation which are 0.00375 and 0.00178125, respectively.

6.2 *Analytical Results of MCLT*

Strains are analytically obtained using MCLT and modified constant. The mathematical results are defined as:

$$\begin{Bmatrix} \varepsilon_{11} \\ \varepsilon_{22} \\ \gamma_{12} \end{Bmatrix} = \begin{bmatrix} \frac{1}{38.8} & \frac{-0.25}{38.8} & 0 \\ \frac{-0.25}{38.8} & \frac{1}{5.406} & 0 \\ 0 & 0 & \frac{1}{3.077} \end{bmatrix} \begin{Bmatrix} 30 \\ 0 \\ 0 \end{Bmatrix} 10^{-3} = \begin{Bmatrix} 0.0007731 \\ -0.0001936 \\ 0 \end{Bmatrix} \quad (17)$$

$$\text{or}\, \varepsilon_{22} = -0.0001936$$

Similarly, strains ε_{22} are obtained for 0° and 45° fiber orientation which are 0.005549 and 0.002871, respectively.

6.3 *Finite Element Analysis of CLT*

Figures 12 and 13a present the strain distribution and strain–time plot for 0°, 45°, 90° fiber orientations corresponding to CLT. It is observed that, the strain values at a node located in mid-section of the specimen are 0.00541, 0.002121, and −0.0001885 for 0°, 45°, 90° fiber orientations, respectively. The FEA results align with the analytical results as presented in previous sub-sections.

6.4 *Finite Element Analysis of CLT*

Figures 14 and 13b present the strain distribution and strain–time plot for 0°, 45°, 90° fiber orientations when MCLT approach is used for finite element analysis. It is observed that the strain values at a node located in mid-section of the specimen are 0.005767, 0.0032114, and −0.0002201 for 0°, 45°, 90° fiber orientations, respectively. The FEA results also align with the analytical results obtained in analytical MCLT approach described in previous subsection. It is to be noted that the strain value calculated in direction 2 in local coordinate system is along the

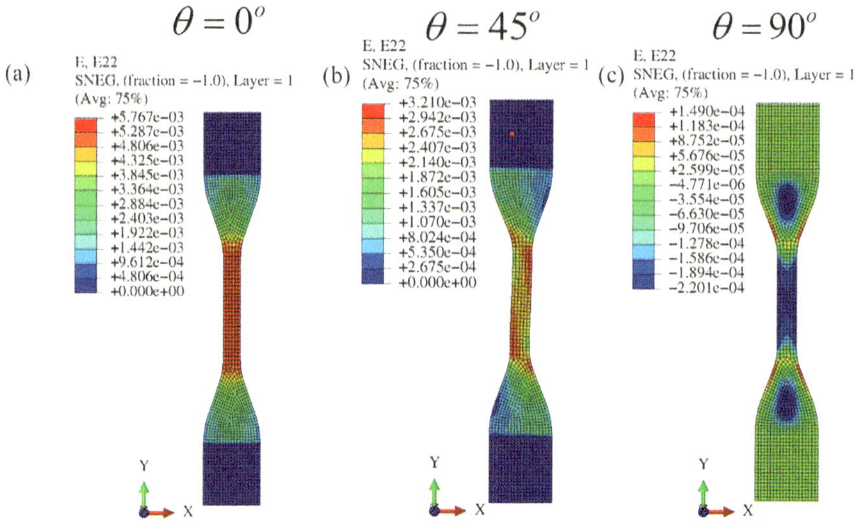

Fig. 12 Strain distributions obtained for: (**a**) 0°, (**b**) 45°, and (**c**) 90° fiber orientations using CLT

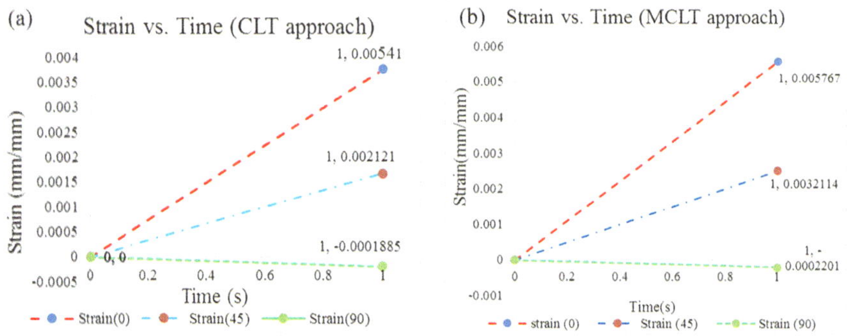

Fig. 13 Strain–time plot for 0°, 45°, 90° fiber orientations using: (**a**) classical laminate theory (CLT) and (**b**) modified classical laminate theory (MCLT)

length of the specimen. It can also be observed that the strain is decreasing as the fiber orientation is changing from 0° to 90° for both CLT and MCLT approaches. Strain values obtained in case of 0°, 45°, and 90° orientation using MCLT approach is greater than those obtained through CLT approach. This is due to the fact that MCLT approach is more realistic as it considers the effect of gaps and voids between the fibers and layers, and thus, it is possible that load-bearing strength is reduced.

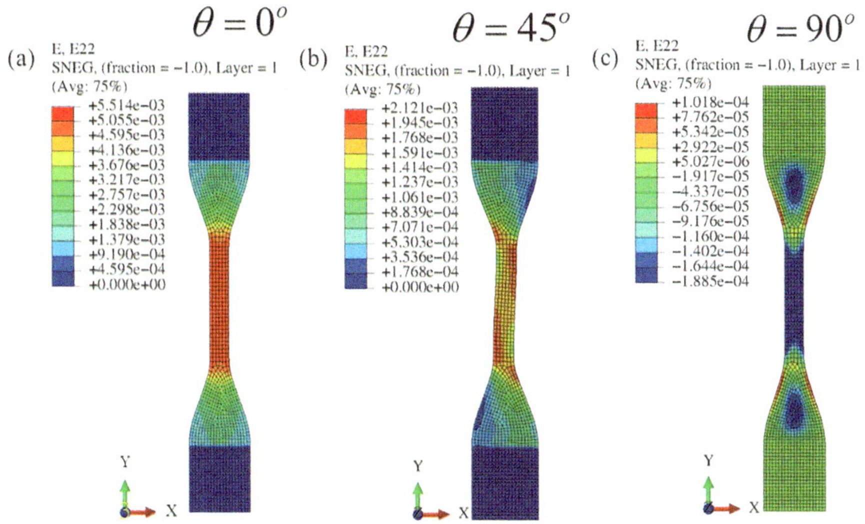

Fig. 14 Strain distributions obtained for: (**a**) 0°, (**b**) 45°, and (**c**) 90° fiber orientations using MCLT

7 Conclusions

This book chapter presents macro-mechanical modeling approaches, namely classical laminate theory (CLT) and modified classical laminate theory (CLT) of composites to observe its suitability in predicting the mechanical behavior of parts manufactured by fused deposition modeling (FDM). The mechanical properties of the polylactic acid (PLA) material used in FDM were considered to simulate the mechanical behavior of the FDM part for three deposition strategy (i.e., 0°, 45°, 90° fiber orientations). An increase in strain values for MCLT approach is observed. Thus, there is an essential need of considering gaps and voids in the CLT theory which assumes that material is deposited without any void. The present study will help the researcher in the selection of a suitable approach to simulate the mechanical behavior of 3D printed parts through fused-deposition modeling.

References

1. Hodder, I. (2014). The entanglements of humans and things: A long-term view. *New Literary History, 45*, 19–36.
2. Gunasekaran, A. (1999). Agile manufacturing: A framework for research and development. *International Journal of Production Economics, 62*, 87–105.
3. Huang, Y., Leu, M. C., Mazumder, J., & Donmez, A. (2015). Additive manufacturing: Current state, future potential, gaps and needs, and recommendations. *Journal of Manufacturing Science and Engineering, 137*(1), 014001.

4. Guide, V. D. R., Jr. (2000). Production planning and control for remanufacturing: Industry practice and research needs. *Journal of Operations Management, 18*, 467–483.
5. Gibson, I., Rosen, D. W., & Stucker, B. (2014). *Others: Additive manufacturing technologies*. Springer.
6. Ligon, S. C., Liska, R., Stampfl, J., Gurr, M., & Mülhaupt, R. (2017). Polymers for 3D printing and customized additive manufacturing. *Chemical Reviews, 117*, 10212–10290.
7. Kiran, A. S. K., Veluru, J. B., Merum, S., Radhamani, A., Doble, M., Kumar, T. S., & Ramakrishna, S. (2018). Additive manufacturing technologies: An overview of challenges and perspective of using electrospraying. *Nanocomposites, 4*, 190–214.
8. Kumar, P., Ahuja, I., & Singh, R. (2012). Application of fusion deposition modelling for rapid investment casting–a review. *International Journal of Materials Engineering Innovation., 3*, 204–227.
9. Ahn, S., Wright, P. K., Montero, M., Odell, D., & Roundy, S. (2002). Anisotropic material properties of fused deposition modeling ABS. *Rapid Prototyping Journal, 8*, 248–257.
10. Pande, S., & S., Kumar, S. (2008). A generative process planning system for parts produced by rapid prototyping. *International Journal of Production Research, 46*, 6431–6460.
11. Schmidt, L., Schricker, K., Bergmann, J. P., Hussenöder, F., & Eiber, M. (2019). Characterization of a granulate-based strand deposition process in the FLM-method for definition of material-dependent process strategies. *Rapid Prototyping Journal, 25*(1), 104–116.
12. Kumar, Y. R. (2012). An application of Taguchi's technique to improve the accuracy of rapid prototyped FDM parts. *International Journal of Materials Engineering Innovation, 3*, 228–246.
13. Jap, N. S., Pearce, G. M., Hellier, A. K., Russell, N., Parr, W. C., & Walsh, W. R. (2019). The effect of raster orientation on the static and fatigue properties of filament deposited ABS polymer. *International Journal of Fatigue, 124*, 328–337.
14. Somireddy, M., Czekanski, A., & Singh, C. V. (2018). Development of constitutive material model of 3D printed structure via FDM. *Materials Today Communications, 15*, 143–152.
15. Tronvoll, S. A., Welo, T., & Elverum, C. W. (2018). The effects of voids on structural properties of fused deposition modelled parts: A probabilistic approach. *The International Journal of Advanced Manufacturing Technology, 97*, 3607–3618.
16. Comb, J. W., Priedeman, W. R., & Turley, P. W. (1994). Layered manufacturing control parameters and material selection criteria. In *The 1994 international mechanical engineering congress and exposition* (pp. 547–549).
17. Mohan, N., Senthil, P., Vinodh, S., & Jayanth, N. (2017). A review on composite materials and process parameters optimisation for the fused deposition modelling process. *Virtual and Physical Prototyping, 12*, 47–59.
18. Bellini, A., & Güçeri, S. (2003). Mechanical characterization of parts fabricated using fused deposition modeling. *Rapid Prototyping Journal, 9*(4), 252–264.
19. Rodríguez, J. F., Thomas, J. P., & Renaud, J. E. (2001). Mechanical behavior of acrylonitrile butadiene styrene (ABS) fused deposition materials. Experimental investigation. *Rapid Prototyping Journal, 7*(3), 148–158.
20. Kollar, L. P., & Springer, G. S. (2003). *Mechanics of composite structures*. Cambridge University Press.
21. Casavola, C., Cazzato, A., Moramarco, V., & Pappalettere, C. (2016). Orthotropic mechanical properties of fused deposition modelling parts described by classical laminate theory. *Materials and Design, 90*, 453–458.
22. Chen, L., Tao, X., & Choy, C. (1999). Mechanical analysis of 3-D braided composites by the finite multiphase element method. *Composites Science and Technology, 59*, 2383–2391.
23. ABAQUS version 6.14. (2018). *Dassault system*. Simulia Corp.

Exploration on Doped TiO_2 Nanostructures for Application in Optoelectronics and Additive Manufacturing of Functional Prototypes

Pawan Kumar and Ranvijay Kumar

1 Introduction

Semiconductor materials are one of the most studied materials in the last few decades due to it specific application in devices, which play an important role to provide the comfort in human life. The electrical conductivity and optical properties of the semiconductor can be tuned using suitable dopant ions. The conductivity and band gap value of semiconductor lie in between conductor and insulator. At zero kelvin, semiconductor behaves as insulator, and at higher temperature, its conductivity increases due to increase in charge carriers in conduction band. The doping of cation with high valency leads towards the increase in the number of electrons in the host lattice and increase in the conductivity of system. The band gap of semiconductor generally decreases due to extra electron trap in the energy level below the conduction band. After exploring all possible way to improve the efficiency of semiconductor-based devices using dopant ions, it was realised that an alternative way should be used to improve the performance of devices. The doping of foreign element in semiconductor material has its own limitations. Tuning the morphology of semiconductor has emerged as a most promising way to improve the interaction mechanism and devise efficiency without any dopant ions. It has been observed that semiconductor nanostructure shows significant change in their property just by changing morphology. The most common semiconductor morphology is shown in Fig. 1.

P. Kumar
Department of Physics, University Institute of Sciences, Chandigarh University, Gharuan, Punjab, India

R. Kumar (✉)
Department of Mechanical Engineering, University Centre for Research and Development, Chandigarh University, Gharuan, Punjab, India

C. Prakash et al. (eds.), *Additive, Subtractive, and Hybrid Technologies*, Mechanical Engineering Series, https://doi.org/10.1007/978-3-030-99569-0_12

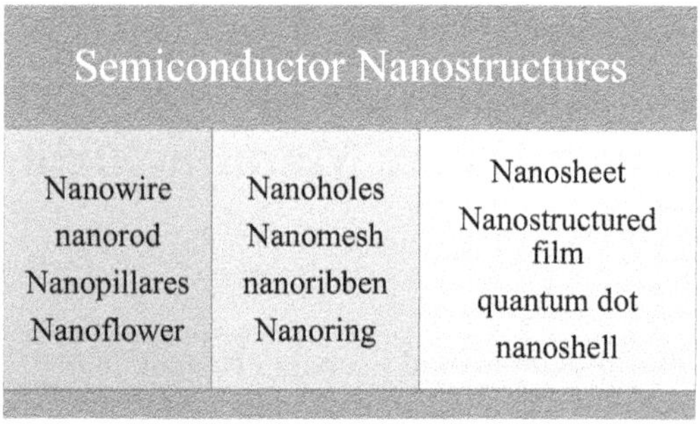

Fig. 1 Different types of possible semiconductor nanostructures

Among all morphology, nanowires have been explored a lot due to its better sensing mechanism. Nanowires are considered as one-dimensional material, which has needle like structure. In case of one-dimensional materials, one dimension is confined in microscale, and other two dimensions are in nanoscale. The nanowire has length of wire in microscale, and the other two dimensions are confined in nanoscale. The most common and useful semiconductors are ZnO, TiO_2 and SnO_2 due to unique properties. The semiconductor-based number of sensors, detector, solar cell and other optoelectronics device has been a part of our daily life. The studies of ZnO nanowire have been reported for gas sensor applications [1]. The TiO_2 nanowires have been reported by several authors [2].

2 Semiconductor (TiO_2) Nanowires

The semiconductor nanostructures have recommendable properties due to change in surface to value ratio. The surface to volume ratio in case of nanostructure is higher than bulk material. The basic region for enhanced sensing mechanism of semiconductor nanowires is the availability of large surface for the reaction. The most explored ZnO, TiO_2 and SnO_2 nanowires report for various applications, and synthesis routes have been summarised in Table 1.

Table 1 shows that TiO_2 nanowire which has been synthesised by various methods for different applications. The doping of transition metal and rare earth has also been employed to explore specific properties.

3 Applications of TiO_2 Nanowire

Although it is difficult to summarise the application of TiO_2 which included various optoelectronics, spintronics and memory devices. Fig. 2 shows the prominent applications of TiO_2 nanowire. The TiO_2 nanowire-based solar cell has been reported by various authors. The study suggests that self-assembled and TiO_2 nanowire can be a promising material to achieve the better photoelectric conversion efficiency. The

Table 1 TiO_2 nanowires and its optoelectronic applications

S. No.	Material functionalities	Synthesis routes	Applications/change in properties	Reference
1	TiO_2 nanowire array	Deposition on substrate using autoclave	Dye-sensitised solar cell	[3]
2	CdS nanorod coated TiO_2 nanowire	Chemical vapour deposition method	Photovoltaic application	[4]
3	N-doped anatase TiO_2 nanowire	Hydrothermal condition	Quantum dot sensitised solar cell	[5]
4	TiO_2 nanowire	Combined hydrothermal and layer by layer assembly	Electrochromic application	[6]
5	TiO_2 nanowire	Sol gel nanostructured	Gas sensor	[7]
6	TiO_2 nanowire	Hydrothermal method	Tuning of photoluminescence property	[8]
7	Composite of TiO_2 nanowire and Nafion	Hydrothermal method	Humidity sensor	[9]
8	Transition metal-doped TiO_2 nanowire	Molten salt flux scheme	Photovoltaic and photocatalysis applications	[10]
9	Pd-doped TiO_2 nanowire	Thermal evaporation method	Hydrogen gas sensors	[11]
10	Cobalt-doped TiO_2	Solvothermal synthesis	Enhances photocatalytic property	[12]
11	N and S co-doped TiO_2 nanowire	Hydrothermal rout	Photocatalytic activity increases for atrazine	[13]
12	Lanthanum and cobalt co-doped TiO_2 nanowire	Solvothermal process	High conductivity for single nanowire field effect transistor	[13]
13	Mn-doped anatase TiO_2 nanowire	Hydrothermal method	TiO_2 and MnO_2 lattice co-exist and show antiferromagnetic and ferromagnetic nature	[14]
14	Fe:TiO_2 nanowire	Spray pyrolysis method	Ethanol sensor	[15]
15	Fe-doped TiO_2 nanowire on graphene	Sol gel method	Enhanced photocatalytic property	[16]
16	Eu + 3-doped TiO_2 nanowire	Hydrothermal techniques	Better photocatalytic activity for methyl orange due to Eu doping	[17]
17	Eu + 3 and Ce+3-doped TiO_2 nanowire arrays	Hydrothermal method	Doped sample shows better decomposition of organic toluidine blue-O dye	[18]

(continued)

Table 1 (continued)

S. No.	Material functionalities	Synthesis routes	Applications/change in properties	Reference
18	Holmium-doped TiO_2 nanowire	Hydrothermal method	Change in structural and photocatalytic properties due to Ho	[19]
19	Si-doped TiO_2 nanowire	Atmospheric pressure chemical vapour deposition method	Improved hydrophilicity	[20]
20	Self (Ti3+)-doped TiO_2 nanowire	Modified hydrothermal method	Enhance absorption in visible region, better photocatalytic activity for methyl orange	[21]

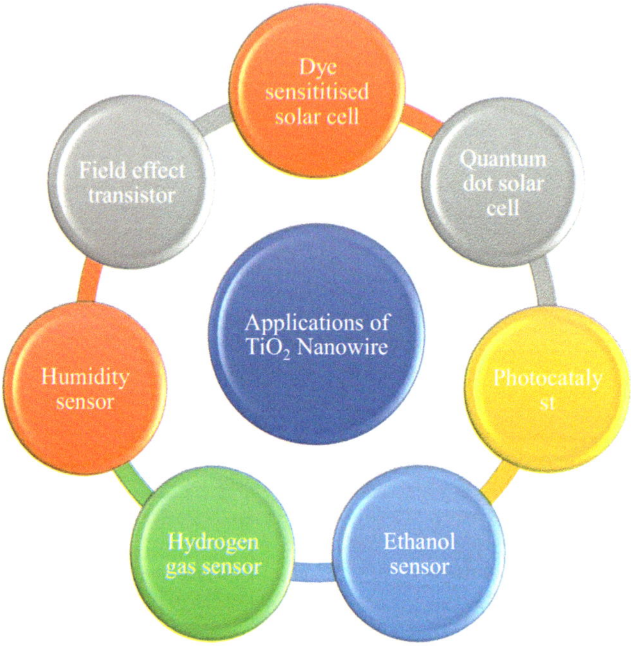

Fig. 2 Applications of TiO_2 nanowires

reported work author has synthesised the nanowire by solvothermal method. Fluorine-doped Tin oxide (FTO) has been used as substrate to be deposited the nanowire array. The photo conversion efficiency reached up to 1.81% using nanowire. Liu et al. have also suggested the importance of TiO_2 nanowire for dye-sensitised solar cell [22]. The author has prepared fibre shaped stretchable TiO_2 nanowire array with the help of solvothermal reaction. The report confirms the 97% photoelectric conversion efficiency by stretching up to 100%. W. C. Tian et al. reported gas sensing behaviour of TiO_2 nanowire [23]. The electron beam lithography techniques have been used to develop the ordered TiO_2 nanowire on p-type Si wafer. The study reveals the fact that ethanol sensing mechanism is independent of ethanol concentration and corelated with operating temperature. G. Wang et al. observed one-dimensional TiO_2-B nanowire as better humidity sensor [24]. The sample has been synthesised by solvothermal method by using TiO_2 which is composed of 30% rutile and 70% anatase. The authors have employed Mott Schottky measurement and electric field-affected photocurrent spectra for analysing the sensing performance of the prepared sample. They revealed the fact that TiO_2-B nanowire has potential application in surface-type humidity sensor. W. Wang et al. have reported an interesting result for TiO_2 nanowire, which may have application in lithium ion-based micro-batteries [25]. The authors have employed template-based electrode deposition for the synthesis of three-dimensional Ni/TiO_2 nanowire network. The result reveals the fact that areal discharging capacity strongly depends on the length of nanowire and increases with an increase in TiO_2 nanowire length. The result of the study suggests that after 600 cycles, the 100% capacity can be retained. In addition, TiO_2 nanostructure has applications in waste management and pollution control devices.

4 TiO_2 Crystal Structure

Titanium dioxide generally exhibits three different crystal structures, namely anatase, rutile and brookite. The unique properties of TiO_2 such as high chemical stability, high refractive index and non-toxicity make it suitable candidate for specific purpose. Anatase phase has tetragonal lattice structure with space group 141/amd. The rutile TiO_2 also exhibits tetragonal lattice but space group is P42/mnm, while brookite-type TiO_2 has orthorhombic lattice with space group Pbca. Formation of different phase depends on the synthesis temperature. If the annealing temperature is lower (below 600°C), the probability of anatase phase is very high while heating the sample above 600°C leads towards the formation of rutile phase TiO_2 nanostructure.

5 Synthesis of Semiconductor Nanowires

A number of techniques have been developed to fabricate the nanowire of specific dimension with high accuracy. Figure 3 summarises the effective techniques for nanowire fabrication of all semiconductor materials. It indicates the effective methods of synthesis based on the available literature [26]. The most effective method for the synthesis of nanowire is hydrothermal techniques, which has been employed by several scientific groups. Here we take example of the TiO_2 nanowire synthesis reported by Zang et al. [8]. In the typical synthesis process, the specific amount of TiO_2 has been placed in a Teflon autoclave chamber. The NaOH solution was inserted in the autoclave to fill 80% volume of the chamber. The temperature was

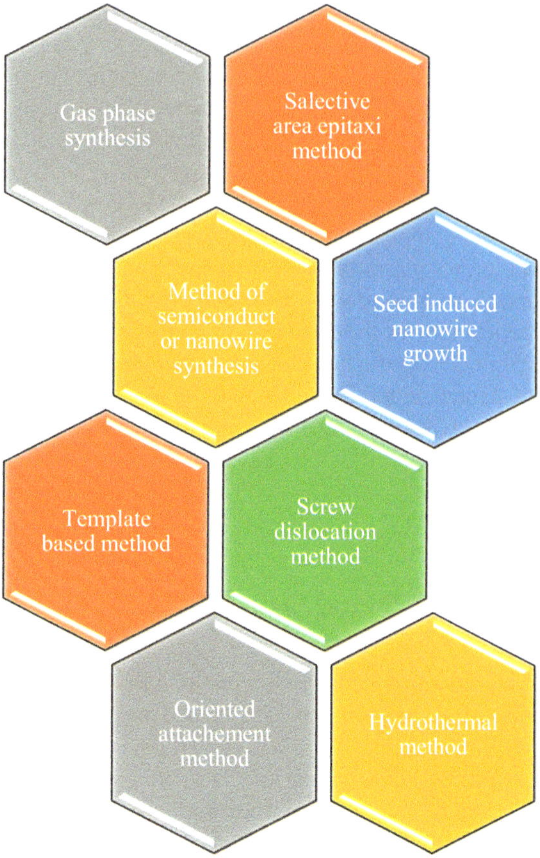

Fig. 3 Synthesis method of semiconductor nanowire fabrications

maintained at 200 °C throughout the process for 1 day. The natural cooling of autoclave provides the sample, which was washed with HCl, distilled water and ethanol to remove all the impurities. The final product was heated at 70 °C for 6 h to get the TiO_2 nanowire.

6 Synthesis of TiO_2 Nanoparticles

TiO_2 nanoparticles can be synthesised by sol gel [27], coprecipitation [28] and hydrothermal method [29] with uniform size and spherical morphology. Behnajady et al. have reported the synthesis of TiO_2 nanoparticles by sol gel method with the help of different starting materials and solvent [30]. The most common precursor used as source of Ti cations are titanium tetra isopropoxide and titanium n-butoxide. The sol gel synthesis of sample significantly affected by different type of solvent and temperature of reaction. The sintering temperature of sample can also tune the size of nanoparticles. It was generally observed that enhancement of sintering temperature leads towards an increase in crystallite size of nanoparticles. Different research group have explored different morphology of TiO_2 for their specific use in optoelectronic application. The sensor and detector application of TiO_2 generally have use of nanowire and nanorod morphology. The basic reason to enhance performance using such morphology is an increase of surface to volume ratio which ultimately provides large surface for reaction.

7 TiO_2 Composite and Its Applications

TiO_2 nanocomposites have been employed for the improvement of TiO_2-based devices. Zhuang has prepared TiO_2 PLA composite by tuning doping concentration of TiO_2. The finding indicates that TiO_2 PLA composite shows better bacteriostatic activity than pure TiO_2 nanoparticles [31]. Table 2 indicates the use of different TiO_2 and polymer composite for different applications.

Table 2 confirms that TiO_2 is suitable for making composite with different polymers for emerging applications. The concentration TiO_2 and selection of polymer will decide the properties of composite. The 3D printing of TiO_2-reinforced polymer has reported as resin for dental denture [47].

8 Framework for Additive Manufacturing of TiO_2-Reinforced Functional Prototypes

The TiO_2-based nanostructures have been emerged as one of the most crucial additive or reinforcement materials in the 3D printing applications. The reinforcement of TiO_2 in thermoplastic-based materials enables the tuning of the mechanical, thermal, structural, chemical and sensing properties which is very much needed in the preparation of functional prototypes. A number of studies have been reported for the preparation of functional prototype using TiO_2-based structures in 3D printing applications. For example, Chen et al. (2019) have reported the 3D printing by using TiO_2 and poly ether ether ketone (PEEK) reinforcement in poly methyl methacryate (PMMA) for the preparation of composite materials to be applied in the preparation of dental applications [47]. Singh et al. have used TiO_2-reinforced PA6 composite feedstock filament for the preparation of functional prototypes in innovative structure manufacturing [48]. Similarly, the 3D printing process has been used in the preparation of dense and porous structures [49], photocatalytic filters [50], photocatalytic thin film [51], hierarchical mesoporous cellular architecture [52] and hierarchical gyroid structures [53]. So, previous researchers have reported tremendously for the manufacturing of functional prototypes using 3D printing processes.

Table 2 Use of different TiO_2 and polymer composite for different applications

S. No.	Composition	Synthesis method	Important change in property/application	Reference
1	TiO_2-PLA thin film	Spin coating	Humidity sensing	[32]
2	PLA-TiO_2	Solvent casting followed by a hot-pressing step	Antibacterial packaging	[33]
3	PLA-modified TiO_2 thin film	Solution blending method	Brittleness of the films increases with nanoparticle concentration	[34]
4	PLA-TiO_2 nanofibre	Electrospinning	Application in antitumor or anticancer drug	[35]
5	PLA-TiO_2 flex fibre	Sol gel dip coating method	Improvement in hygroscopic mechanical and physical properties	[36]
6	PLA film containing TiO_2 and Ag	Solvent evaporation method	Better retention in firmness, colour, and Vit C	[37]
7	Nano-Ag-graphene-TiO_2-PLA film.	Extrusion	Mechanical resistance of the film improved by 30%, water vapour permeability decreased	[38]
8	PVDF/TiO_2 hybrid microfiltration membranes	Thermally induced phase separation method	Change in pure water flux, contact angle, porosity, tensile strength and elongation	[39]
9	PVDF membrane with dispersed TiO_2	Phase inversion method	Modification in membrane surface	[40]
10	Fluorosilane coated TiO_2 into the PVDF-HFP	Electrospinning	Higher water vapour flux	[41]
11	PVDF-TiO_2 membranes	NIPS wet process	Water treatment	[42]
12	PVDF-TiO_2 membranes	TiO_2 sol gel method and blending method	Enhanced membrane strength and thermal stability	[43]
13	TiO_2-ABS nanocomposite	Chemical method	3D printing	[44]
14	TiO_2-ABS	HPh experiment	Electroless and electrolytic deposits show high adherence	[45]
15	TiO_2-ABS polymer	Extracted from Elix polymers	UV-laser making	[46]

Fused deposition modelling (FDM) or in general fused filament fabrication (FFF) is one of the low cost or economical additive manufacturing processes which require the thermoplastic-based materials for the preparation of functional prototypes. The reinforcement of the micro- or nano-sized materials enables its properties to be significantly tuned on the thermal, electrical, thermal sensing properties viewpoint. The reinforcement of the TiO_2-based nanostructures will significantly tune the properties of thermoplastic materials when reinforced which is very crucial in mechanical, biomedical, medical, biotechnology, aerospace, industrial and household applications. So, for this reason, a framework has been presented in this study for the manufacturing of functional goods by using TiO_2 as the reinforcement materials in thermoplastic functional prototypes by 3D printing process (see Fig. 4). The preparation of TiO_2-based functional prototypes can be managed by the selection of the method of the TiO_2 nanostructures, this may be co-precipitation, sol-gel, electro-spinning and hydrothermal methods. In the next step, the synthesised nanoparticles can be mechanical mixed to ensure the uniform dispersion of nanoparticles with thermoplastics. The mechanical mixed thermoplastic granules and TiO_2-based nanostructures are now ready to be extruded with the twin-screw extrusion process for the development of innovative feedstock filaments. The mechanical properties of the prepared feedstock filaments are greatly dependent upon the controlling parameters such as rate of extrusion, screw rotational speed, torque, load applied, etc. Further, the optimised set of process parameters may be employed for the mass manufacturing of TiO_2-based feedstock filaments, and then, those feedstock filaments may be used for the 3D printing process of functional structures/ prototypes.

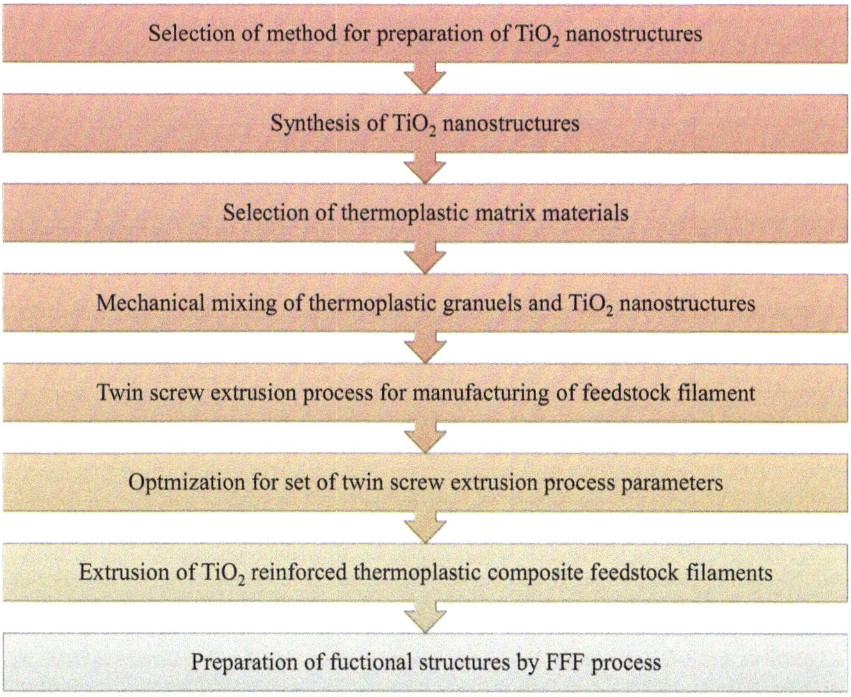

Fig. 4 Framework for preparation of TiO_2-based functional prototypes by 3D printing

9 Summary

In the present work, we have summarised the importance of TiO_2 nanowire for opto-electronics and additive manufacturing. It has been discussed the semiconductors and its properties along with various possible morphology.

It has been observed that various synthesis route has been adopted by different groups for precise fabrication of TiO_2 nanowire. The literature reported in table also indicated that various transition metal and rare earth material had been chosen as suitable candidate for enhanced performance of TiO_2-based sensors and other opto-electronic devices.

The crystal structure of TiO_2 has been described in the next section followed by the description of hydrothermal synthesis method. TiO_2 and polymer composite work summarised has been confirming the utilisation of TiO_2-Polymer composite for prospective material for 3D printing.

The reinforcement of the TiO_2 based nanostructures will significantly tune the properties of thermoplastic materials when reinforced which is very crucial in mechanical, biomedical, medical, biotechnology, aerospace, industrial and household applications. So, for this reason, a framework has been present in this study for the manufacturing of functional goods by using TiO_2 as the reinforcement materials in thermoplastic functional prototypes by 3D printing process.

Acknowledgement The authors sincerely acknowledge the department of Physics UIS Chandigarh University for providing technical facilities to conduct the research work.

References

1. Wan, Q., Li, Q. H., Chen, Y. J., Wang, T. H., He, X. L., Li, J. P., & Lin, C. L. (2004). Fabrication and ethanol sensing characteristics of ZnO nanowire gas sensors. *Applied Physics Letters, 84*(18), 3654–3656.
2. Wang, G., Wang, H., Ling, Y., Tang, Y., Yang, X., Fitzmorris, R. C., Wang, C., Zhang, J. Z., & Li, Y. (2011). Hydrogen-treated TiO2 nanowire arrays for photoelectrochemical water splitting. *Nano Letters, 11*(7), 3026–3033.
3. Kumar, A., Madaria, A. R., & Zhou, C. (2010). Growth of aligned single-crystalline rutile TiO2 nanowires on arbitrary substrates and their application in dye-sensitized solar cells. *The Journal of Physical Chemistry C, 114*(17), 7787–7792.
4. Lee, J. C., Kim, T. G., Lee, W., Han, S. H., & Sung, Y. M. (2009). Growth of CdS nanorod-coated TiO2 nanowires on conductive glass for photovoltaic applications. *Crystal Growth & Design, 9*(10), 4519–4523.
5. Qiu, Q., Li, S., Jiang, J., Wang, D., Lin, Y., & Xie, T. (2017). Improved electron transfer between TiO2 and FTO interface by N-doped anatase TiO2 nanowires and its applications in quantum dot-sensitized solar cells. *The Journal of Physical Chemistry C, 121*(39), 21560–21570.
6. Liu, S., & Qu, X. (2017). Construction of nanocomposite film of Dawson-type polyoxometalate and TiO2 nanowires for electrochromic applications. *Applied Surface Science, 412*, 189–195.
7. Francioso, L., Taurino, A. M., Forleo, A., & Siciliano, P. (2008). TiO2 nanowires array fabrication and gas sensing properties. *Sensors and Actuators B: Chemical, 130*(1), 70–76.

8. Zhang, Y. X., Li, G. H., Jin, Y. X., Zhang, Y., Zhang, J., & Zhang, L. D. (2002). Hydrothermal synthesis and photoluminescence of TiO2 nanowires. *Chemical Physics Letters, 365*(3–4), 300–304.
9. Wu, R. J., Sun, Y. L., Lin, C. C., Chen, H. W., & Chavali, M. (2006). Composite of TiO2 nanowires and Nafion as humidity sensor material. *Sensors and Actuators B: Chemical, 115*(1), 198–204.
10. Liu, B., Chen, H. M., Liu, C., Andrews, S. C., Hahn, C., & Yang, P. (2013). Large-scale synthesis of transition-metal-doped TiO2 nanowires with controllable overpotential. *Journal of the American Chemical Society, 135*(27), 9995–9998.
11. Hoa, N. T. Q. (2015). Solvothermal synthesis and photocatalytic activity of Co–doped TiO2 nanowires. *International Journal of Nanotechnology, 12*(5–7), 426–433.
12. Zhang, Y., Liu, P., & Wu, H. (2015). Development of high efficient visible light-driven N, S-codoped TiO2 nanowires photocatalysts. *Applied Surface Science, 328*, 335–343.
13. Li, S., Wang, Z., Yang, Y., Li, J., & Jin, C. (2020). Enhanced conductivity of anatase TiO2 nanowires by La and co co-doping. *Micro & Nano Letters, 15*(4), 226–229.
14. Li, H. M., Liu, M., Zeng, Y. S., & Huang, T. C. (2010). Coexistence of antiferromagnetic and ferromagnetic in Mn-doped anatase TiO 2 nanowires. *Journal of Central South University of Technology, 17*(2), 239–243.
15. Vijayalakshmi, K., & Jereil, S. D. (2015). Enhanced ethanol sensing performance of Fe: TiO2 nanowires and their mechanism of sensing at room temperature. *Ceramics International, 41*(2), 3220–3226.
16. Farhangi, N., Chowdhury, R. R., Medina-Gonzalez, Y., Ray, M. B., & Charpentier, P. A. (2011). Visible light active Fe doped TiO2 nanowires grown on graphene using supercritical CO2. *Applied Catalysis B: Environmental, 110*, 25–32.
17. Zhao, S. Q., Wan, H. L., & Asuha, S. (2013). Preparation and properties of Eu3+–doped TiO2 nanowires. In *Advanced materials research* (Vol. 624, pp. 67–71). Trans Tech Publications Ltd.
18. Bandi, V. R., Murugesan, R. C., Grandhe, B. K., Kim, S. S., Jang, K., Shin, D.-S., Yi, S.-S., & Jeong, J.-H. (2013). Synthesis, structural and optical properties of pure and rare-earth ion doped TiO2 nanowire arrays by a facile hydrothermal technique. *Thin Solid Films, 547*, 207–211.
19. Zhou, W., & He, Y. (2012). Ho/TiO2 nanowires heterogeneous catalyst with enhanced photocatalytic properties by hydrothermal synthesis method. *Chemical Engineering Journal, 179*, 412–416.
20. Du, J., Li, X., Li, K., Gu, X., Qi, W., & Zhang, K. (2016). High hydrophilic Si-doped TiO2 nanowires by chemical vapor deposition. *Journal of Alloys and Compounds, 687*, 893–897.
21. Wang, J., Yang, P., & Huang, B. (2015). Self-doped TiO2– x nanowires with enhanced photocatalytic activity: Facile synthesis and effects of the Ti3+. *Applied Surface Science, 356*, 391–398.
22. Liu, G., Wang, H., Wang, M., Liu, W., Ardhi, R. E. A., Zou, D., & Lee, J. K. (2018). Study on a stretchable, fiber-shaped, and TiO2 nanowire array-based dye-sensitized solar cell with electrochemical impedance spectroscopy method. *Electrochimica Acta, 267*, 34–40.
23. Tian, W. C., Ho, Y. H., Chen, C. H., & Kuo, C. Y. (2013). Sensing performance of precisely ordered TiO2 nanowire gas sensors fabricated by electron-beam lithography. *Sensors, 13*(1), 865–874.
24. Wang, G., Wang, Q., Lu, W., & Li, J. (2006). Photoelectrochemical study on charge transfer properties of TiO2– B nanowires with an application as humidity sensors. *The Journal of Physical Chemistry B, 110*(43), 22029–22034.
25. Wang, W., Tian, M., Abdulagatov, A., George, S. M., Lee, Y. C., & Yang, R. (2012). Three-dimensional Ni/TiO2 nanowire network for high areal capacity lithium ion microbattery applications. *Nano Letters, 12*(2), 655–660.
26. Dasgupta, N. P., Sun, J., Liu, C., Brittman, S., Andrews, S. C., Lim, J., Gao, H., Yan, R., & Yang, P. (2014). 25th anniversary article: Semiconductor nanowires–synthesis, characterization, and applications. *Advanced Materials, 26*(14), 2137–2184.

27. Sharma, A., Karn, R. K., & Pandiyan, S. K. (2014). Synthesis of TiO2 nanoparticles by sol-gel method and their characterization. *Journal of Basic and Applied Engineering Research, 1*(9), 1–5.
28. Sanchez-Martinez, A., Ceballos-Sanchez, O., Koop-Santa, C., López-Mena, E. R., Orozco-Guareño, E., & García-Guaderrama, M. (2018). N-doped TiO2 nanoparticles obtained by a facile coprecipitation method at low temperature. *Ceramics International, 44*(5), 5273–5283.
29. Hidalgo, M. C., Aguilar, M., Maicu, M., Navío, J. A., & Colón, G. (2007). Hydrothermal preparation of highly photoactive TiO2 nanoparticles. *Catalysis Today, 129*(1–2), 50–58.
30. Behnajady, M. A., Eskandarloo, H., Modirshahla, N., & Shokri, M. (2011). Investigation of the effect of sol–gel synthesis variables on structural and photocatalytic properties of TiO2 nanoparticles. *Desalination, 278*(1–3), 10–17.
31. Zhuang, W., Liu, J., Zhang, J. H., Hu, B. X., & Shen, J. (2009). Preparation, characterization, and properties of TiO2/PLA nanocomposites by in situ polymerization. *Polymer Composites, 30*(8), 1074–1080.
32. Mallick, S., Ahmad, Z., Touati, F., Bhadra, J., Shakoor, R. A., & Al-Thani, N. J. (2018). PLA-TiO2 nanocomposites: Thermal, morphological, structural, and humidity sensing properties. *Ceramics International, 44*(14), 16507–16513.
33. Segura González, E. A., Olmos, D., Lorente, M. Á., Vélaz, I., & González-Benito, J. (2018). Preparation and characterization of polymer composite materials based on PLA/TiO2 for antibacterial packaging. *Polymers, 10*(12), 1365.
34. Buzarovska, A. (2013). PLA nanocomposites with functionalized TiO2 nanoparticles. *Polymer-Plastics Technology and Engineering, 52*(3), 280–286.
35. Song, M., Pan, C., Li, J., Wang, X., & Gu, Z. (2006). Electrochemical study on synergistic effect of the blending of nano TiO2 and PLA polymer on the interaction of antitumor drug with DNA. *Electroanalysis: An International Journal Devoted to Fundamental and Practical Aspects of Electroanalysis, 18*(19–20), 1995–2000.
36. Foruzanmehr, M., Vuillaume, P. Y., Elkoun, S., & Robert, M. (2016). Physical and mechanical properties of PLA composites reinforced by TiO2 grafted flax fibers. *Materials & Design, 106*, 295–304.
37. Chi, H., Song, S., Luo, M., Zhang, C., Li, W., Li, L., & Qin, Y. (2019). Effect of PLA nano-composite films containing bergamot essential oil, TiO2 nanoparticles, and Ag nanoparticles on shelf life of mangoes. *Scientia Horticulturae, 249*, 192–198.
38. Peter, A., Cozmuta, L. M., Nicula, C., Cozmuta, A. M., Talasman, C. M., Drazic, G., Peñas, A., Calahorro, A. J., Sagratini, G., & Silvi, S. (2021). Chemical and organoleptic changes of curd cheese stored in new and reused active packaging systems made of Ag-graphene-TiO2-PLA. *Food Chemistry, 363*, 130341.
39. Shi, F., Ma, Y., Ma, J., Wang, P., & Sun, W. (2012). Preparation and characterization of PVDF/TiO2 hybrid membranes with different dosage of nano-TiO2. *Journal of Membrane Science, 389*, 522–531.
40. Oh, S. J., Kim, N., & Lee, Y. T. (2009). Preparation and characterization of PVDF/TiO2 organic–inorganic composite membranes for fouling resistance improvement. *Journal of Membrane Science, 345*(1–2), 13–20.
41. Lee, E. J., An, A. K., Hadi, P., Lee, S., Woo, Y. C., & Shon, H. K. (2017). Advanced multi-nozzle electrospun functionalized titanium dioxide/polyvinylidene fluoride-co-hexafluoropropylene (TiO2/PVDF-HFP) composite membranes for direct contact membrane distillation. *Journal of Membrane Science, 524*, 712–720.
42. Méricq, J. P., Mendret, J., Brosillon, S., & Faur, C. J. C. E. S. (2015). High performance PVDF-TiO2 membranes for water treatment. *Chemical Engineering Science, 123*, 283–291.
43. Yu, L. Y., Shen, H. M., & Xu, Z. L. (2009). PVDF–TiO2 composite hollow fiber ultrafiltration membranes prepared by TiO2 sol–gel method and blending method. *Journal of Applied Polymer Science, 113*(3), 1763–1772.

44. Skorski, M. R., Esenther, J. M., Ahmed, Z., Miller, A. E., & Hartings, M. R. (2016). The chemical, mechanical, and physical properties of 3D printed materials composed of TiO2-ABS nanocomposites. *Science and Technology of Advanced Materials, 17*(1), 89–97.
45. Magallón-Cacho, L., Pérez-Bueno, J. J., Meas-Vong, Y., Stremsdoerfer, G., & Espinoza-Beltrán, F. J. (2011). Surface modification of acrylonitrile-butadiene-styrene (ABS) with heterogeneous photocatalysis (TiO2) for the substitution of the etching stage in the electroless process. *Surface and Coatings Technology, 206*(6), 1410–1415.
46. Clemente, M. J., Lavieja, C., Peña, J. I., & Oriol, L. (2018). UV-laser marking of a TiO2-containing ABS material. *Polymer Engineering & Science, 58*(9), 1604–1609.
47. Chen, S. G., Yang, J., Jia, Y. G., Lu, B., & Ren, L. (2019). TiO2 and PEEK reinforced 3D printing PMMA composite resin for dental denture base applications. *Nanomaterials, 9*(7), 1049.
48. Singh, R., Kumar, R., Mascolo, I., & Modano, M. (2018). On the applicability of composite PA6-TiO2 filaments for the rapid prototyping of innovative materials and structures. *Composites Part B: Engineering., 143*, 132–140.
49. Aleni, A. H., Kretzschmar, N., Jansson, A., Ituarte, I. F., & St-Pierre, L. (2020). 3D printing of dense and porous TiO2 structures. *Ceramics International, 46*(10), 16725–16732.
50. Sangiorgi, A., Gonzalez, Z., Ferrandez-Montero, A., Yus, J., Sanchez-Herencia, A. J., Galassi, C., Sanson, A., & Ferrari, B. (2019). 3D printing of photocatalytic filters using a biopolymer to immobilize TiO2 nanoparticles. *Journal of the Electrochemical Society, 166*(5), H3239.
51. Do, H. H., Tran, T. K., Ung, T. D., Dao, N. T., Nguyen, D. D., Trinh, T. H., Hoang, T. D., Le, T. L., & Tran, T. T. (2021). Controllable fabrication of photocatalytic TiO2 brookite thin film by 3D-printing approach for dyes decomposition. *Journal of Water Process Engineering., 1*(43), 102319.
52. Torres Arango, M. A., Kwakye-Ackah, D., Agarwal, S., Gupta, R. K., & Sierros, K. A. (2017). Environmentally friendly engineering and three-dimensional printing of TiO2 hierarchical mesoporous cellular architectures. *ACS Sustainable Chemistry & Engineering., 5*(11), 10421–10429.
53. Jo, W., Yoon, B. J., Lee, H., & Moon, M. W. (2017). 3D printed hierarchical gyroid structure with embedded photocatalyst TiO2 nanoparticles. *3D Printing and Additive Manufacturing, 4*(4), 222–230.
54. Wei, Z., Yao, Y., Huang, T., & Yu, A. (2011). Solvothermal growth of well-aligned TiO2 nanowire arrays for dye-sensitized solar cell: Dependence of morphology and vertical orientation upon substrate pretreatment. *International Journal of Electrochemical Science, 6*(6), 1871–1879.

Welding-Based Additive Manufacturing for Biomedical Applications: From Concept to Technology

Pawan Sharma, Gurminder Singh, B. N. Sahoo, Girish C. Verma, Pulak M. Pandey, and Vishvesh J. Badheka

1 Introduction

The demand for complex and customized shape products has been increasing continuously in the area of biomedical engineering, especially in orthopaedic and cardiovascular applications. However, the fabrication of aforesaid products is challenging and requires advanced manufacturing technologies. Currently, additive manufacturing (AM) processes, i.e. stereolithography (SL), fused deposition modelling (FDM), selective laser sintering/melting (SLS/SLM), binder jet 3D printing, electron beam melting (EBM), laminated objective manufacturing (LOM) and laser engineered net shaping (LENS), have been used for the manufacturing of such products [1]. However, only limited AM techniques such as SLM, EBM, LOM, LENS, etc. are used for the fabrication of metal parts. Furthermore, limitations such as

P. Sharma (✉) · B. N. Sahoo
Department of Mechanical Engineering, Sardar Vallabhbhai National Institute of Technology, Surat, Gujarat, India
e-mail: pawansharma@med.svnit.ac.in

G. Singh
SIMAP Lab, Université Grenoble Alpes, Gières, France

G. C. Verma
Department of Mechanical Engineering, Indian Institute of Technology Indore, Indore, Madhya Pradesh, India

P. M. Pandey
Department of Mechanical Engineering, Indian Institute of Technology Delhi, New Delhi, India

V. J. Badheka
School of Technology, Mechanical Engineering Department, Pandit Deendayal Petroleum University, Gandhinagar, Gujarat, India

C. Prakash et al. (eds.), *Additive, Subtractive, and Hybrid Technologies*, Mechanical Engineering Series, https://doi.org/10.1007/978-3-030-99569-0_13

slower manufacturing speeds when compared to conventional manufacturing, difficulty in process control, lack of compatible working material, substrate requirement, the requirement of post-process finishing and high capital cost of the machine as well as tooling restrict the application of existing AM techniques in the manufacturing of complex metal parts. Researchers are working continuously to improve the technology in AM to overcome the aforementioned limitations. However, the initial capital cost of the above-mentioned AM technologies for metal part fabrications is still on a higher side, and high capital cost investment is required for the establishment of these technologies. Hence, for increasing the applicability of AM for the manufacturing of complex and customized shape biomedical products, there exists a need for the development of a low-cost AM system with improved process characteristics.

Over the past few years, welding-based additive manufacturing (WBAM) processes exhibited promising characteristics for the fabrication of customized and large-scale metal components. The most popularly used WBAM technique is wire and arc AM (WAAM). Generally, in WAAM, an electric arc is used as the heat source and a solid material wire as the feedstock. The WBAM systems offer various advantages over the existing AM system for metal fabrication such as high deposition rate, relatively low building time, and low investment cost [2]. Different arc welding techniques, such as gas metal arc welding (GMAW), gas tungsten arc welding (GTAW), and plasma arc welding (PAW), have been employed to fabricate metal parts [3]. However, limitations such as high energy input, residual stresses in fabricated metal parts, and poor part finishing are associated with the aforementioned WAAM processes. Recently, with extension to AM with arc, researchers introduced a new power source with different waveform designs, known as cold metal transfer (CMT). CMT offers a unique advantage of a higher deposition rate with lower heat input as compared to other arc-based AM techniques. Hence, it could be hypothesized that the CMT process may be effectively used for the development of a low-cost AM technique for the manufacturing of complex-shaped parts at a faster speed and improved accuracy. Currently, WBAM systems are extensively researched by both academia and industry owing to their several advantages. Therefore, it is important to understand the fundamental concept and principles of WBAM technology. Hence, this chapter aims to provide an insight into the major developments in the WBAM systems. Details related to the technological concept, process variants, materials, and product characterization will be discussed. Furthermore, current applications based on WBAM and the future scope of currently developed technology will be discussed.

2 Welding-Based Additive Manufacturing (WBAM) Systems

As per the ASTM F2792-12a [4], additive manufacturing is defined as a process wherein layer by layer joining of materials is obtained to make objects from 3D model data. The raw material used in AM can be in the form of solid wire, powder,

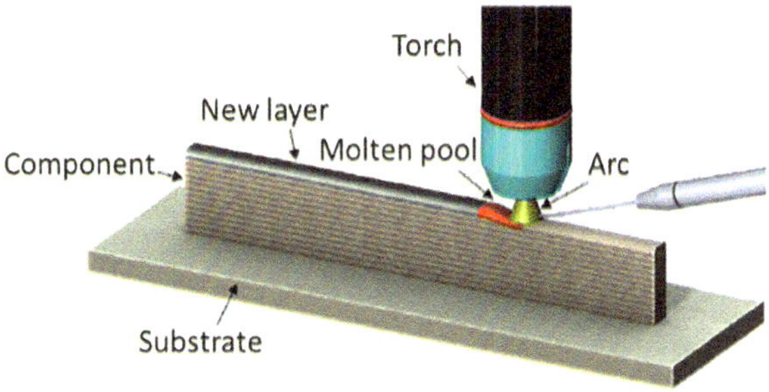

Fig. 1 Schematic WAAM process [5]

sheet, and liquid. In the case of metal objects, generally, metal powders or wires are being utilized for the end part manufacturing. Furthermore, the fusion of materials is achieved using high intensity energy sources such as a laser beam, electron beam, etc. Therefore, to facilitate proper layer by layer fusion of materials, a high intensity energy source is required. Hence, an electric arc in welding can be used as an effective heat source to be utilized as an AM system. The material is then fed in the form of wire feedstock. The aforesaid discussed WBAM process is commonly known as wire and arc-based AM (WAAM). Figure 1 shows the schematic of the WAAM process. According to ASTM F2792-12a [4], the WAAM process is classified as a type of direct energy deposition (DED) AM process. Thus, an electric arc acts as a source of direct energy or heat source and metal wire as a feedstock material.

2.1 *Benefits of WBAM Techniques*

Electrical arc as a source for achieving fusion between layers offers numerous processing benefits as compared to the other sources used in DED processes such as electron beam and laser [6]. Benefits of the WBAM process as compared to other DED fusion sources are listed as below:

1. Low capital investment, owing to the easy availability of open-source welding equipment [7].
2. Higher wall plug efficiencies as compared to laser beam or electron beam technologies [8].
3. Ease in processing conditions such as no need for a vacuum environment to operate as necessary for electron beam-based DED technique [8].
4. As compared to laser-based DED technique, the electrical arc as a fusion source offers a higher efficiency [9]. The higher energy efficiency of an electric arc is beneficial especially for reflective metal and metal alloys having poor laser coupling efficiency such as copper [10], aluminium, and magnesium [11].
5. The high rate of deposition with the capability to process a wide range of materials.

2.2 Robotic WBAM System

Robotic WBAM system consists of a source of power, an articulated robotic system for work table control or motion mechanism control (generally computer numerically controlled), an automatic wire feedstock material system, and necessary accessories for achieving better process control such as shielding gas, the system for preheating, and cooling system.

A schematic of a robotic WBAM system has been shown in Fig. 2 [12]. It consists of a computer interface, a robotic controller, a programmable welding power source, and an industrial robotic manipulator. The computer interface is required for programming the process and for collecting the experimental results for further analysis. The function of the robot controller includes controlling the coordination of both the robot motions and the welding processes. The welding process is controlled using a programmable welding power source. Subsequently, an industrial robot manipulator helps in implementing the movement of the torch for metal deposition. Finally, a structure as per the designed strategy is deposited as shown in Fig. 2.

Generally, in WBAM systems, two different types of designs are used. The first design consists of a completely closed chamber for attaining inert gas shielding environment, similar to laser-based DED systems. The second design involves the use of an existing or specifically designed mechanism of gas shielding, with the robot located on a linear rail to enhance the overall working envelope. Using this type of design, large metallic structures having dimensions varying from several meters could be fabricated easily. Figure 3 depicts an example of such a design as used for the robotic WBAM system.

2.3 Fabrication Process Using WBAM System

Fabrication of a metal part using the WBAM system involves mainly three steps, i.e. process planning, deposition of metal, and post-processing. As per the designed CAD model, pre-processing such as 3D slicing and generation, the desired motion

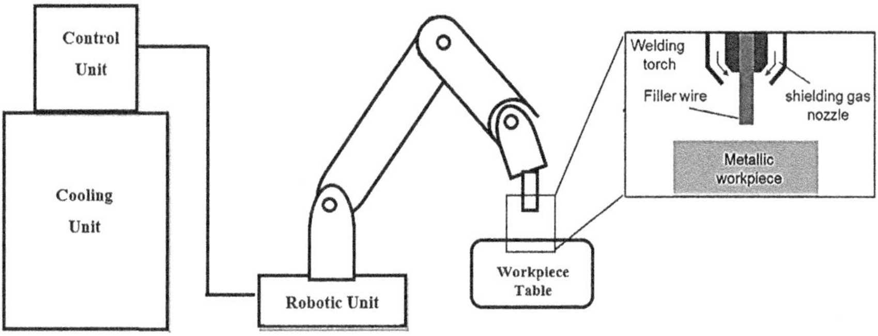

Fig. 2 Experimental setup of the WBAM system

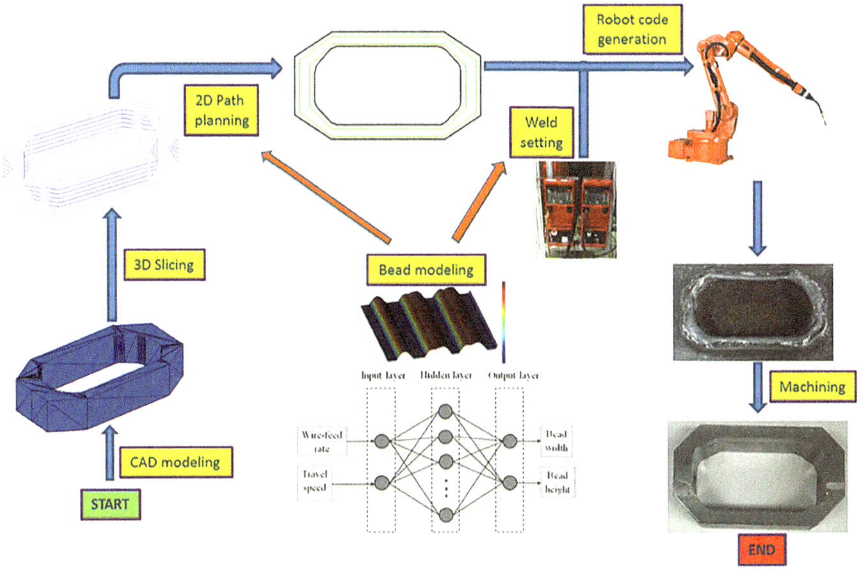

Fig. 3 Experimental set up of robotic welding-based system [13]

of the robotic arm as well as process parameters for the deposition is achieved using an appropriate interfacing software. This step aims to produce a defect-free part with the desired geometrical accuracy [14, 15]. Based on the model of deposition to be used in welding for the specific part, the 3D slicing and interfacing software for programming allow for automated planning of tool path and optimization of process parameters to avoid any processing defects [13, 16]. During the fabrication step, the robot and external axis mechanism allow accurate welding torch motion to facilitate the layer-by-layer building of a component. An example of the metal parts fabricated using the WBAM system illustrating the steps involved is shown in Fig. 3. Furthermore, an advanced WBAM system contains different sensors for the measurement of key parameters such as welding signals, metal transfer behaviour, deposited bead geometry, and interpass temperature [17]. Hence, such systems support in situ process monitoring and better control to achieve a higher quality product.

3 Variants of WBAM Process

Various research groups over the world, working on different techniques of WBAM systems, are tabulated in Table 1.

More broadly depending on the fundamental concepts of automatized welding processes or the source of arc, WBAM systems are classified as gas metal arc welding (GMAW), gas tungsten arc welding (GTAW), and plasma arc welding (PAW) [2].

Table 1 Different research groups over the world working on WBAM

Nomenclature for WBAM	Institute/university	Energy source	Reference
Three-dimensional (3D) welding	University of Nottingham	GMAW	[18]
Near-net shape manufacturing	Tufts University	GMAW, PAW	[19]
3D welding	Korea Institute of Science and Technology	GMAW	[9, 10]
Welding-based rapid prototyping	University of Kentucky	GMAW	[11, 12]
GMAW-based rapid prototyping	Harbin Institute of Technology	GMAW	[20]
Welding-based deposition	Southern Methodist University	GMAW, GTAW	[7, 8]
Shape deposition manufacturing	Carnegie Mellon and Stanford University	GMAW, GTAW	[21]
Hybrid-layered manufacturing	IIT Hyderabad and IIT Bombay	GMAW	[22]
MPAW-based rapid prototyping	Xi'an Jiaotong University	Micro-PAW	[23]
WAAM	University of Wollongong	GMAW, GTAW	[24–26]
AM	Cranfield University	GMAW, GTAW, PAW	[18, 19, 27]

3.1 GMAW-Based AM System

GMAW also referred to as metal inert gas (MIG) welding is a process wherein a fusion between the metal layers is obtained owing to the energy of an electric arc formed between the tip of a consumable wire electrode and the workpiece material. In the case of the GMAW-based AM system, the metal wire is fed continuously and is transferred through the electric arc into the resulting molten pool. The complete fusion process occurs under the shielding of inert gas to allow complete protection of the weld pool and adjacent layer of material. Deposition rates varying from 15 to 160 *g/min* can be obtained using GMAW-based AM process [28–30]. Therefore, depending on the processing conditions such as the type of material to be deposited and process parameters, GMAW-based AM systems allow for the production of large-scale metal parts in a short period. However, limitations such as higher heat input and residual stresses in the final part restrict the application of the GMAW-based AM process. Thus, to achieve controlled deposition of molten metal with reduced heat input, a different variant of GMAW, i.e. cold metal transfer (CMT) and tandem GMAW, has been developed by the researchers [31].

Cold Metal Transfers GMAW AM System

CMT-based AM system is an advanced technology which works on the principle of controlled dip transfer mode mechanism. It is a modified GMAW process wherein the material transfer is obtained owing to short-circuiting transfer. A control system in the CMT process immediately detects the short-circuiting just as the tip of electrode wire makes contacts with the molten pool, thereby sending signals for the activation of a servomotor. Subsequently, the electrode wire is retracted using a push and pull electromechanical system promoting droplet transfer of metal as shown in Fig. 4 [32]. Different variants of CMT such as CMT pulse, CMT advanced, and CMT pulse advanced are available for use in AM [33]. It was reported that the deposition porosity in the fabricated part has been significantly affected by the type of arc mode in the CMT process. Conventional CMT was found to be inappropriate for the AM owing to the existence of a high quantity of gas pores in the deposited layer [33]. On the other hand, excellent performance in terms of controlling porosity was obtained in CMT pulse advanced technique, thereby making it the most suitable process for AM applications.

Tandem GMAW AM System

Tandem GMAW-based AM system employs a twin-wire configuration wherein two wires are deposited simultaneously to achieve a high rate of deposition as shown in Fig. 5 [34]. Furthermore, to increase the rate of deposition and efficiency of the system, another variant of GMAW wherein a double-electrode GMAW (DE-GMAW) system utilizing GTAW torch to offer bypass current has also been developed as shown in Fig. 6 [35]. An increase in the coefficient of material utilization of more

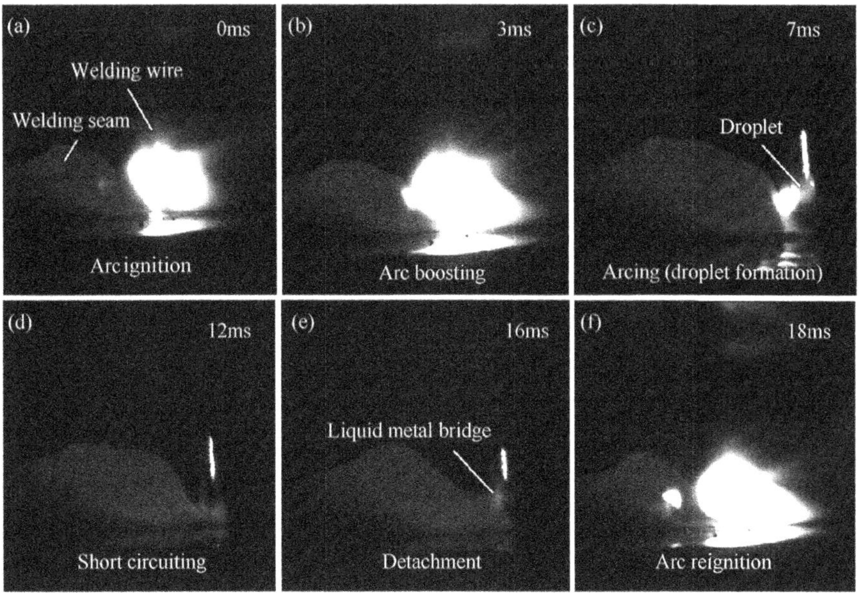

Fig. 4 Mechanism of droplet transfer in CMT [32]

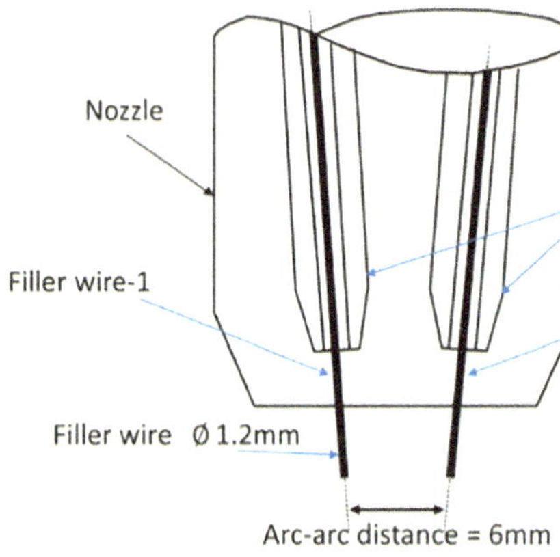

Fig. 5 Schematic of twin-wire welding torch for the WBAM system [34]

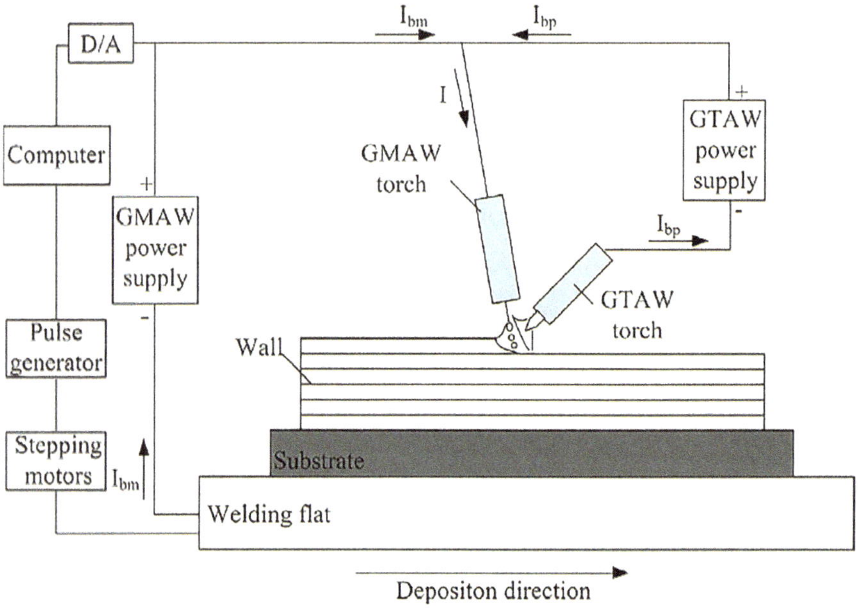

Fig. 6 Schematic diagram of the double-electrode GMAW-based AM system [35]

than 10% was reported when depositing thin-walled parts using DE-GMAW within a suitable range of bypass current. However, the limitation of the aforesaid discussed variants of GMAW is that they require high energy for maintaining the electric arc. Therefore, such processes need further modifications to achieve better control of the molten pool shape through enhanced heat dissipation.

3.2 GTAW-Based AM System

GTAW-based AM systems employ a non-consumable tungsten electrode which is used to establish an electric arc between the electrode and the workpiece under the environment of inert shielding gas. As compared to GMAW-based AM system, GTAW-based AM systems are characterized by better arc control with high precision and fewer defects [36]. The process planning in the case of the GTAW-based AM system is more complicated owing to the variability in the wire feed orientation which subsequently affects the quality of the deposit. The orientation of wire feed largely affects material transfer and the deposit quality. Generally, used feeding methods in GTAW-based AM are back, side, and front feeding, respectively [37]. Normally for AM of Ti-based and Fe-based materials, the front feeding method is used.

3.3 PAW-Based AM Systems

PAW-based AM systems work on a principle of a high energy density deposition wherein the electric arc is forced to travel through a constricted orifice located between the anode and the cathode. The system for micro PAW-based AM has been shown in Fig. 7 [23]. Hence, better arc stability as compared to GMAW and GTAW is obtained in this system. In PAW-based AM system, a highly localized channel of ionized plasma is formed containing very high thermal energy. The energy densities in the PAW-AM system are lower as compared to high power beams such as laser but higher than GMAW-AM and GTAW-AM system [38]. Improved control in the deposited weld bead geometry is obtained in this process owing to the existence of a narrow and thin heat-affected zone. This control can be achieved by independently controlling the important process variables such as current and wire feed [39]. Furthermore, different operating modes can be achieved in PAW based AM systems through the variation of parameters such as torch orifice diameter, plasma gas flow rate, and current intensity. Different modes in PAW-based AM systems are [31]:

1. Microplasma
2. Medium current
3. Keyhole plasma

1. *Microplasma*: This mode is obtained using low welding currents, usually ranging from 0.1 and 15 A. Pertaining to AM, this mode of PAW offers advantages related to the total wall width of the fabricated sample. It allows for the production of thin-walled parts having total wall width values as low as 2 *mm*. The rate of deposition obtained in this method is approximately 1.0 *g/min* [40]. Compared to other WBAM systems, the major benefit of this method is the requirement of low heat input which allows for a better surface finish.

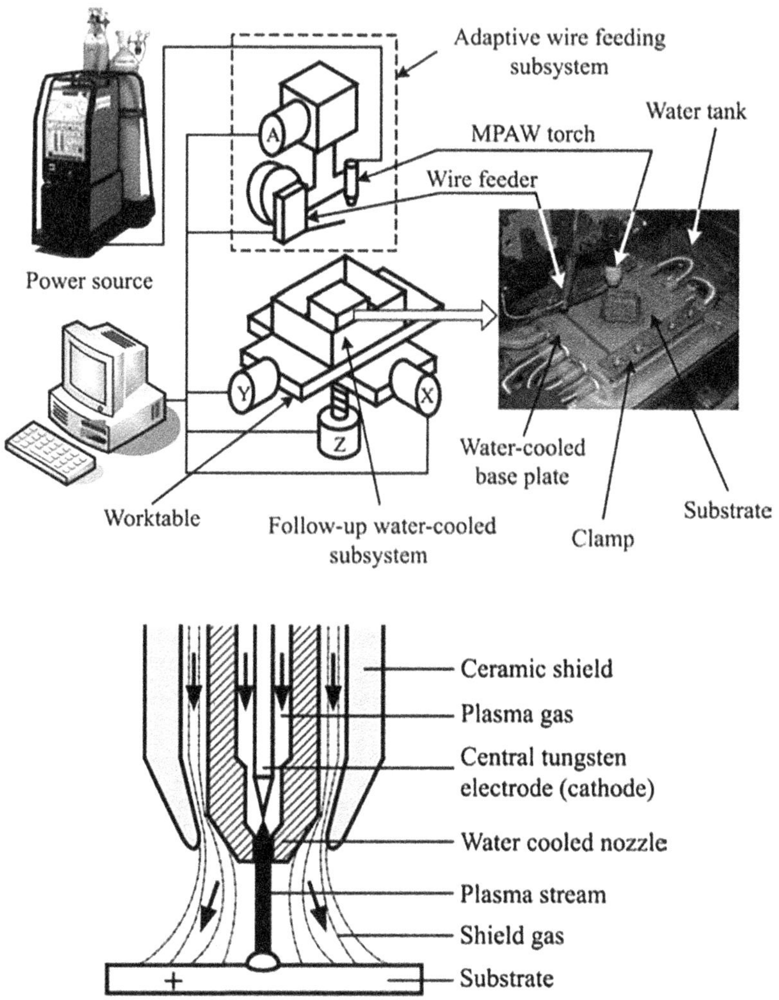

Fig. 7 Schematic diagram of the micro PAW-based AM system and plasma arc details [23]

2. *Medium current*: In this mode of PAW, medium current values are used, usually ranging from 15 to 200 *A*. This is the most commonly used mode of PAW-based AM system. This operation mode has characteristics quite similar to GTAW-based AM system; however, the better arc control is obtained in this case. Total wall width varying from 4 to 15 *mm* and the deposition rate as high as 30 *g/min* can be obtained using this mode in PAW-based AM system [28, 41, 42].
3. *Keyhole mode*: This mode in PAW is characterized by high current values resulting in very high penetration. Therefore, owing to such a high penetration, this mode is generally unsuitable for AM purpose as it may lead to the melting of the previously deposited layers, thereby affecting the wall stability and geometric accuracy [31].

3.4 Solid State Additive Manufacturing

Solid state additive manufacturing (SSAM) utilizes solid state welding processes like ultrasonic welding, friction welding, friction stir welding, etc. to overcome the problems associated with the aforesaid discussed WBAM processes. These include problems such as contamination, delamination and cracks, residual stresses, and porosity in the final part. As there is no melting of the metal, there is no possibility of any solidification defects like porosities, inclusions, delamination, etc. taking place. These processes can also be used for the creation of functionally graded structures where various dissimilar metal combinations can be utilized. Also due to having a lower number of process parameters, the parameters can be optimized for efficiency for numerous combinations.

The two major classifications in SSAM are ultrasonic additive manufacturing and friction-based additive manufacturing. Friction-based AM further contains processes like linear friction AM, rotary friction AM, friction stir AM, additive friction stir AM, friction surfacing AM, and friction stir processing AM.

4 Methodologies Used for Quality Improvement in the WBAM System

Various methodologies have been applied to improve material properties, reduce porosity and surface roughness, and eliminate residual stresses and distortions in the part. These methodologies have been discussed in this section in brief detail.

4.1 Cold Metalworking-Based Method

A cold working method such as cold rolling has been found to be effective in improving the quality of the parts fabricated using WBAM. In this technique, plastic deformation using a high load (up to 100 kN) has been obtained on the weld bead joining each layer, thereby reducing residual stresses and distortion [43]. This method is referred to as interpass rolling, and the schematic of the process has been shown in Fig. 8 [44]. Along with the reduction in residual stresses and distortion, this method also results in improved material properties owing to the attainment of homogeneous grain refinement. Furthermore, better control in the width of the parts, and improved surface finish can be achieved using this method [45].

Other cold working techniques such as machine hammer peening [46] and laser shock peening [47] have been reported to be efficacious in improving the mechanical properties of welded-based AM-fabricated parts. Improved hardness and grain refinement with the reduction in average grain size by 22% have been achieved using laser shock peening of 2319 AL alloy part. However, this technique was found to be effective for a very small thickness and was not suitable for achieving grain

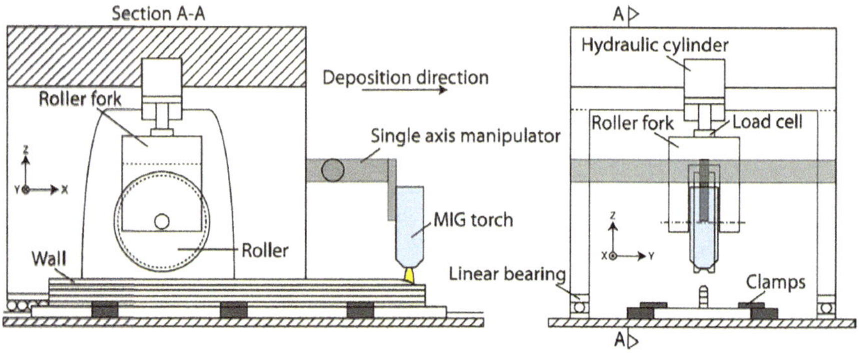

Fig. 8 Schematic diagram of the cold metalworking WBAM system [44]

refinement in wall thickness. Though cold metalworking methods were efficacious in improving microstructural homogeneity and mechanical properties, reducing residual stresses, and achieving increased final part accuracy, the involvement of interpassing step and heavy equipment may reduce the overall lead time of the system. Therefore, it may result in a reduction in the rate of deposition in WBAM systems.

4.2 *Post-Process Heat Treatment*

Post-process heat treatment in the final fabricated part is widely adopted in WBAM to decrease residual stress and improve mechanical properties and microstructural properties. The heat treatment process to be selected largely depends on the target material, the WBAM technique, temperature for working, and heat treatment conditions. It has been reported that after the post-process heat treatment of the fabricated part, the mechanical strength of WBAM-fabricated parts improved quite significantly. An increase of 4%, 78%, 5%, and 17% has been reported for Ti alloy, Al alloy, Ni-based superalloys, and intermetallic Ti/Al alloy, respectively [17]. Furthermore, it helps in achieving refinement of grains, as reported for WBAM-fabricated Al and Inconel alloy [48]. Most commonly, materials with high carbon % must be post heat treated, while some of the materials may get damaged owing to this treatment. Hence, careful consideration is necessary before the application of the post-heat treatment process to the WBAM component.

4.3 *Interpass Heat Reduction/Interpass Cooling*

The interlayer temperature (the temperature of the deposited layer immediately before the next layer is deposited) plays a significant role in achieving better part quality in WBAM processes [49]. High heat at the interface and low cooling rates

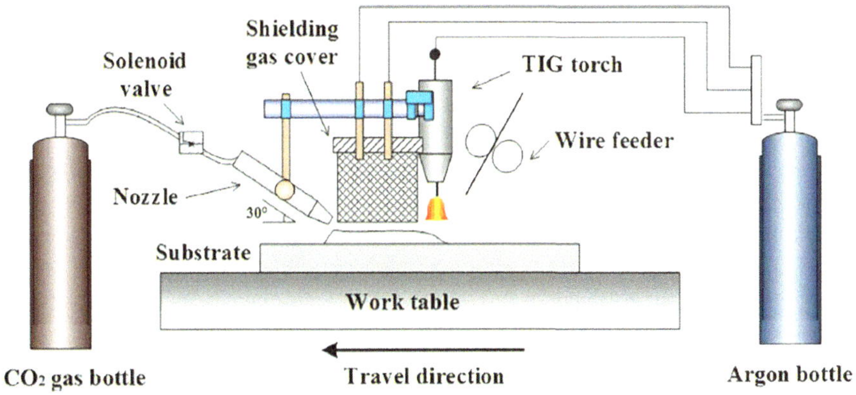

Fig. 9 Gas cooling WBAM system [17]

result in coarse columnar grain structure and subsequently anisotropy in the fabricated part. Different techniques have been employed for controlling the interlayer temperature [31]. In one of the techniques, the inter-layer temperature was controlled using an optimized interlayer idle time evaluated through finite element simulation of the thermal behaviour during part [49].

Another technique to control the temperature is the use of interpass cooling as depicted in Fig. 9. Interpass cooling is obtained by supplying gas such as argon, nitrogen or CO_2 through the moveable gas nozzle. Controlled cooling results in better microstructure and desired mechanical properties. Pertaining to the interpass cooling technique, a study has been performed using compressed CO_2 to fabricate *Ti6Al4V* thin-walled structures [50]. Refined microstructure, improved hardness, higher strength, and low surface oxidation were found owing to interpass cooling. Furthermore, manufacturing efficiency was improved significantly owing to the decrease in dwell time between the deposited layers.

5 Problems and Defects Associated with Welding-Based Additive Manufacturing

5.1 Deformation and Residual Stress

Similar to the powder bed-based AM process, samples fabricated using WBAM techniques are also characterized by the existence of residual stresses and distortion. The residual stresses result in part distortion, lack of geometric tolerance, layers delamination during deposition, weakening of fracture resistance, and inferior fatigue performance. Therefore, control and reduction of residual stress and deformation are important. Different deformation types that occur in WBAM-fabricated parts include shrinkage along the longitudinal and transverse direction, distortion due to bending, angular as well as rotational deformation [17]. The reason for these deformations could be attributed to the thermal expansion and shrinkage in the fabricated parts owing to repetitive melting and cooling of the material. Pertaining to AM of large thin-walled structures, the existence of such a phenomenon is a major concern [44].

Residual stresses can be referred to as the stress which remains in the fabricated part even after the removal of all external loading forces. These stresses significantly affect the mechanical properties and fatigue performance in the fabricated parts. Cracking will occur once the residual stress goes beyond the ultimate tensile strength (UTS) of the material, whereas warping or plastic deformation will occur once it exceeds the yield strength but is less than UTS [51]. Among all WBAM engineering materials, parts fabricated using two different metals, i.e., bimetal parts, show high levels of residual stress and deformation owing to the difference in thermal expansion of the material. Various process parameters such as welding current, voltage, rate of wire feed, temperature, the flow rate of the shielding gas, etc. significantly affect the existence of residual stresses and deformation in the final fabricated part [2, 17].

5.2 *Porosity and Lack of Fusion*

Porosity is a major concern in WBAM parts as it results in low mechanical strength owing to the micro-crack formation. Furthermore, porosity causes a reduction in fatigue strength. Commonly in WBAM, this type of defect is classified mainly as raw material-induced [52] as well as process-induced [53]. The reason for porosity could be attributed to surface contamination due to the presence of moisture, grease, and some other hydrocarbon compounds in the raw material. Absorption of these contaminants in the molten pool results in porosity generation. Most commonly, Al alloys were found to be more susceptible to this flaw owing to the hydrogen solubility in both solid and liquid states. After the solidification, even if the small amount of dissolved hydrogen in the liquid state exceeds the solubility limit, then it may result in porosity [54]. Hence, the purity of raw materials to be used is important, especially for Al alloys. Furthermore, another reason for porosity could be related to process conditions caused mainly due to the unstable deposition or poor path planning process. Such type of porosity is generally non-spherical in shape. Gaps or voids are produced in the affected region owing to the insufficient fusion or spatter that occurred because of the complexity in the deposition path or variations in the manufacturing process.

5.3 *Accuracy and Surface Finish*

In addition to the deformation occurred owing to residual stresses, another significant reason for poor accuracy and low surface finish in a final part is the un-match and the staircase effect that occurs due to improper slicing manner. Generally, the slicing of a solid model of any part is obtained using a constant layer thickness. The un-match occurs when the sliced model with a constant layer thickness does not match with the original solid model. Pertaining to high deposition rate and speeds, the WBAM processes are effective for simpler geometries. However, when high-levels of accuracy in the final parts are desired, generally a milling process is used along with the WBAM technique.

5.4 Cracking and Delamination

Cracking and delamination are other significant defects related to WBAM techniques and occur mainly due to thermal stresses and the characteristics of the deposited material. Generally, cracks in WBAM parts are categorized as grain boundary cracks or solidification cracks [17]. Obstruction in solidified grain flow or high strain in the weld pool is the main cause of the solidification-type crack. Cracking along the grain boundary occurs owing to the variances between morphology at the boundary and formation of precipitate or dissolution. Furthermore, the major cause for delamination or separation between the adjacent layers could be attributed to the incomplete melting or fusion between the layers. Commonly, this defect is visible and cannot be corrected by a post-process treatment. One of the solutions to overcome this defect is the preheating of the substrate.

6 Applications of WBAM Processes for Biomedical Sector

In the present era, AM has become a major medical technology because it can fabricate implants as per the needs of each person like patient-specific implants. Currently, with the world's elderly population rising rapidly, demand for personalized medicine and customizable 3D printed medical equipment, especially surgical implants and orthopaedic tools, are increasing day-by-day rapidly. Thus, the demand for medical devices has been tried to fulfil by metal printed implants. The properties advantage and low price of WBAM over the other AM processes give rise to the fabrication of patient-specific implant at a lower cost. It is possible to fabricate biomedical grade metal materials like stainless steel, titanium alloys, magnesium alloys, and nitinol alloy—the reported investigations of these materials as shown in Fig. 10. The mechanical properties and metallurgical properties of WBAM-fabricated materials have been reported on the higher side as compared to other metal AM processes. Also, WBAM techniques give freedom to alter the metallurgy of the material, which could be helpful to control the corrosion resistance and biocompatibility of the implants. Although most components developed with WBAM technology possess basic geometries, geometric precision is probably not as high as other AM technologies. In the planning of metal surgical equipment systems, complicated geometries are required. The WBAM approach is, however, only in its very

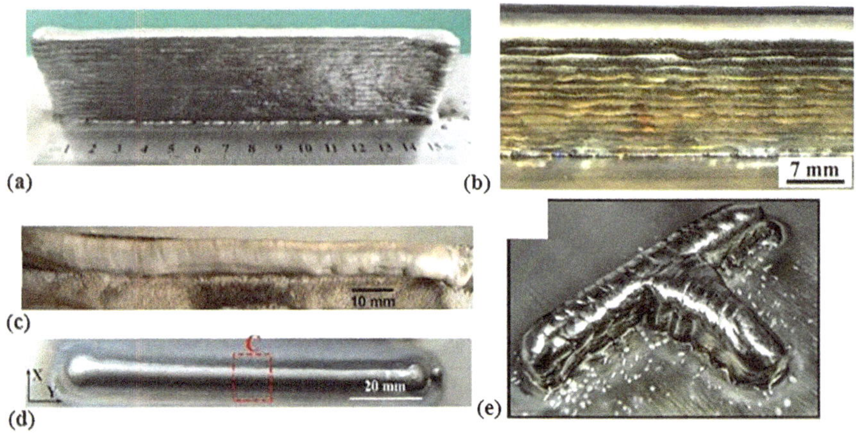

Fig. 10 WAAM of different biomedical-related materials like (**a**) stainless steel [55], (**b**) titanium alloy [50], (**c**) magnesium alloy [11], (**d**) Nitinol [56], and (**e**) thin wall structure of titanium alloy [57]

beginning stage for biomedical uses, and before its therapeutic use is universal, there are many obstacles to control. One of the biggest challenges for the WBAM is to control geometry features. This plays an important role to fabricate patient-specific implants. Moreover, high accuracy post-processing is required after WBAM fabrication to control fine features.

WBAM provides the ability to manufacture topologically optimized materials, as they become very costly, with high material loss and extensive lead times by traditional technologies. This also opens new application area to fabricate topological ordered scaffolds. However, as discussed above, the fabrication of fine geometrical features is still a big challenge. The biomedical industry is primarily focused on the manufacture of sophisticated parts of titanium, stainless steel, and magnesium alloys, making WBAM a cost-effective production process regardless of the difficulties involved with the subtractive processes used with these materials. However, many other bio-metals such as tantalum, cobalt-chromium, or iron have not been explored by this process.

Moreover, there are some other issues for fabricating metal implants with high thermal inputs of WBAM process. High thermal input is responsible for two main problems: residual stress and distortion. By the plastic deformation of the material in WBAM process, residual stress and distortion damage dramatically geometrical uniformity and also impact the mechanical properties and performance of parts. Besides, because of the reduction in the efficient fatigue life and tensile strength of the parts, the lifetime of the manufactured metal components is reduced. The thermal expansion of the heated specimen, which induces compressive stress, restricts the surrounding cold substances during the fusion process. The tensile stress behind the heat source can result from the contraction of cooling substances with nearby cooled materials. Fortunately, with considerable work, counter techniques have been found to reduce or remove residual tension and distortion. Therefore, the control of process parameters is required by the experimental method and even by simulation method to alter the thermal properties to fabricate the desired properties in the material for implant fabrication.

Biomedical grade alloys were used in WBAM but still do not satisfy the criteria for corrosion resistance and biocompatibility in their use for biomedical applications. Also, only few works have been reported for the fabrications of bio-metal by WBAM. The road for the investigations for the biomedical applications has not been explored in-depth. With improvement and development in process for controlling geometry issues and investigating new materials, it would be highly appreciated in the biomedical sector. Some possible applications in the biomedical sector are shown in Fig. 11. The low-cost advantage could be used to fabricate general-purpose complex implants such as hip, knee, shoulder, or oral implants in production with SS 316 and Ti6Al4V. CAD control process also gives the advantage to fabricate patient-specific implants by extracting geometry features from tomography scan of the patient.

The feasibility of fabricating random porosity by shape holder materials or ordered porous implants circumvents the stress shielding effect brought by dense materials due to mismatch of stiffness and elastic modulus as compared to human bone tissue. Furthermore, the surgical tools and in vitro medical devices such as prosthesis, hearing aids, and dental surgery or template models are quite feasible to fabricate by WBAM process with material and process development.

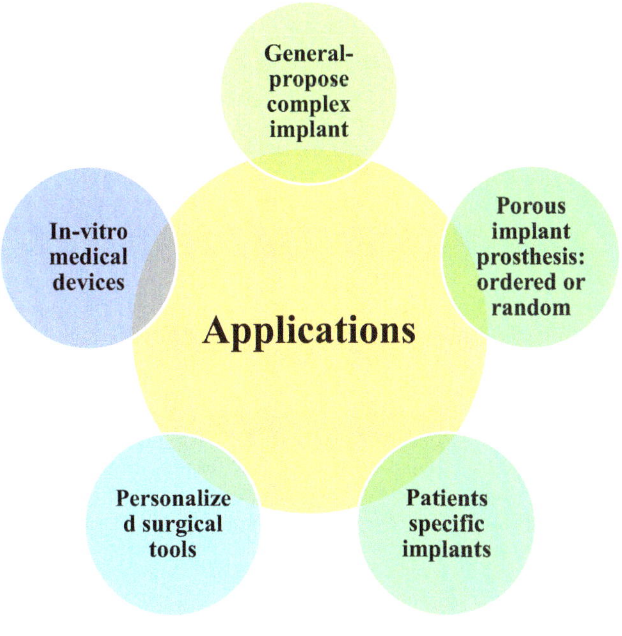

Fig. 11 Applications of WAAM process in the biomedical sector

7 Conclusions and Future Scope

Welding-based AM has been found to possess ideal characteristics as desired for the fabrication of complex and customized shape biomedical parts. High deposition rates, reduced porosity, and low cost of WBAM processes make them an ideal substitute for the existing AM technique for biomedical part fabrication. Research and development of WBAM for the fabrication of metallic parts is interdisciplinary in nature, integrating the concepts of materials science, thermo-mechanical engineering, and process planning. Recently, developments in different WBAM technologies have been done to achieve optimized microstructure and mechanical properties in the final fabricated part. However, further work is necessary to achieve better process control and optimization especially in the direction of online process monitoring and reduction in residual stresses as well as distortions. Most of the biomaterials such as titanium and its alloys, stainless steel, magnesium and its alloys, and nickel-based alloys have been processed successfully using WBAM processes. However, only few dedicated research attempts have been made related to the processing of biomedical parts using WBAM. Hence, comprehensive work in the direction of processing of biomedical parts using WBAM is necessary for the successful implementation of this technology in biomedical area. Furthermore, biocompatibility, mechanical integrity, corrosion behaviour, patient-specific case studies, and in vivo studies of WBAM-fabricated parts need to be addressed as recently only limited work in this direction has been performed.

References

1. Huang, Y., Leu, M. C., Mazumder, J., & Donmez, A. (2015). Additive manufacturing: Current state, future potential, gaps and needs, and recommendations. *Journal of Manufacturing Science and Engineering, 137*, 1–10. https://doi.org/10.1115/1.4028725
2. Ding, D., Pan, Z., Cuiuri, D., & Li, H. (2015). Wire-feed additive manufacturing of metal components: Technologies, developments and future interests. *International Journal of Advanced Manufacturing Technology, 81*, 465–481. https://doi.org/10.1007/s00170-015-7077-3
3. Fang, X., Zhang, L., Li, H., Li, C., Huang, K., & Lu, B. (2018). Microstructure evolution and mechanical behavior of 2219 aluminum alloys additively fabricated by the cold metal transfer process. *Materials (Basel), 11*, 812. https://doi.org/10.3390/ma11050812
4. ASTM International F2792-12a. (2013). *Standard terminology for additive manufacturing technologies*. https://doi.org/10.1520/F2792-12A.2.
5. McAndrew, A. R., Alvarez Rosales, M., Colegrove, P. A., Hönnige, J. R., Ho, A., Fayolle, R., Eyitayo, K., Stan, I., Sukrongpang, P., Crochemore, A., & Pinter, Z. (2018). Interpass rolling of Ti-6Al-4V wire + arc additively manufactured features for microstructural refinement. *Additive Manufacturing, 21*, 340–349. https://doi.org/10.1016/j.addma.2018.03.006
6. Cunningham, C. R., Flynn, J. M., Shokrani, A., Dhokia, V., & Newman, S. T. (2018). Invited review article: Strategies and processes for high quality wire arc additive manufacturing. *Additive Manufacturing, 22*, 672–686. https://doi.org/10.1016/j.addma.2018.06.020
7. Anzalone, G., Zhang, C., Wijnen, B., Sanders, P., Pearce, J., & Zhang, C. (2013). A low-cost open-source metal 3-D printer. *IEEE Access, 1*, 803–810.

8. Milewski, J. O. (2017). Additive manufacturing of metals; from fundamental technology to rocket nozzles. *Medical Implants, and Custom Jewelry*, 85–97. https://doi.org/10.1007/978-3-319-58205-4
9. Jackson, M. A., Van Asten, A., Morrow, J. D., Min, S., & Pfefferkorn, F. E. (2016). A comparison of energy consumption in wire-based and powder-based additive-subtractive manufacturing. *Procedia Manufacturing, 5*, 989–1005. https://doi.org/10.1016/j.promfg.2016.08.087
10. Gu, D. D., Meiners, W., Wissenbach, K., & Poprawe, R. (2012). Laser additive manufacturing of metallic components: Materials, processes and mechanisms. *International Materials Review, 57*, 133–164. https://doi.org/10.1179/1743280411Y.0000000014
11. Guo, J., Zhou, Y., Liu, C., Wu, Q., Chen, X., & Lu, J. (2016). Wire arc additive manufacturing of AZ31 magnesium alloy: Grain refinement by adjusting pulse frequency. *Materials (Basel), 9*. https://doi.org/10.3390/ma9100823
12. Ding, D., Pan, Z., Cuiuri, D., & Li, H. (2014). A tool-path generation strategy for wire and arc additive manufacturing. *International Journal of Advanced Manufacturing Technology, 73*, 173–183. https://doi.org/10.1007/s00170-014-5808-5
13. Ding, D., Pan, Z., Cuiuri, D., Li, H., Van Duin, S., & Larkin, N. (2016). Bead modelling and implementation of adaptive MAT path in wire and arc additive manufacturing. *Robotics and Computer-Integrated Manufacturing, 39*, 32–42. https://doi.org/10.1016/j.rcim.2015.12.004
14. Ding, D., Pan, Z., Cuiuri, D., & Li, H. (2015). A practical path planning methodology for wire and arc additive manufacturing of thin-walled structures. *Robotics and Computer-Integrated Manufacturing, 34*, 8–19. https://doi.org/10.1016/j.rcim.2015.01.003
15. Ding, D., Pan, Z., Cuiuri, D., & Li, H. (2015). A multi-bead overlapping model for robotic wire and arc additive manufacturing (WAAM). *Robotics and Computer-Integrated Manufacturing, 31*, 101–110. https://doi.org/10.1016/j.rcim.2014.08.008
16. Ding, D., Pan, Z., Cuiuri, D., Li, H., & van Duin, S. (2016). Advanced design for additive manufacturing: 3D slicing and 2D path planning. In *New trends 3D print* (pp. 3–24). https://doi.org/10.5772/63042
17. Wu, B., Pan, Z., Ding, D., Cuiuri, D., Li, H., Xu, J., & Norrish, J. (2018). A review of the wire arc additive manufacturing of metals: Properties, defects and quality improvement. *Journal of Manufacturing Processes, 35*, 127–139. https://doi.org/10.1016/j.jmapro.2018.08.001
18. Spencer, J. D., Dickens, P. M., & Wykes, C. M. (1997). Rapid prototyping of metal parts by three-dimensional welding. *Proceedings of the Institution of Mechanical Engineers, Part B, 212*, 175–182.
19. Kwak, Y. M., & Doumanidis, C. C. (2002). Geometry regulation of material deposition in near-net shape manufacturing by thermally scanned welding. *Journal of Manufacturing Processes, 4*, 28–41. https://doi.org/10.1016/S1526-6125(02)70131-X
20. Xiong, J., Zhang, G., Gao, H., & Wu, L. (2013). Modeling of bead section profile and overlapping beads with experimental validation for robotic GMAW-based rapid manufacturing. *Robotics and Computer-Integrated Manufacturing, 29*, 417–423. https://doi.org/10.1016/j.rcim.2012.09.011
21. Neplotnik, G., Ramaswami, K., Padmanabhan, P., Merz, R., Weiss, L. E., Schultz, L., & Prinz, F. B. (1997). Shape deposition manufacturing of heterogeneous structures. *Journal of Manufacturing Systems, 16*, 239–248.
22. Suryakumar, S., Karunakaran, K. P., Bernard, A., Chandrasekhar, U., Raghavender, N., & Sharma, D. (2011). Weld bead modeling and process optimization in hybrid layered manufacturing. *CAD Computer-Aided Design, 43*, 331–344. https://doi.org/10.1016/j.cad.2011.01.006
23. Aiyiti, W., Zhao, W., Lu, B., & Tang, Y. (2006). Investigation of the overlapping parameters of MPAW-based rapid prototyping. *Rapid Prototyping Journal, 12*, 165–172. https://doi.org/10.1108/13552540610670744
24. Song, Y. A., Park, S., & Chae, S. W. (2005). 3D welding and milling: Part II - Optimization of the 3D welding process using an experimental design approach. *International Journal of Machine Tools and Manufacture, 45*, 1063–1069. https://doi.org/10.1016/j.ijmachtools.2004.11.022

25. Zhang, Y. M., Chen, Y., Li, P., & Male, A. T. (2003). Weld deposition-based rapid prototyping: A preliminary study. *Journal of Materials Processing Technology, 135*, 347–357. https://doi.org/10.1016/S0924-0136(02)00867-1
26. Zhang, Y. M., Li, P., Chen, Y., & Male, A. T. (2002). Automated system for welding-based rapid prototyping. *Mechatronics, 12*, 37–53.
27. Song, Y. A., Park, S., Choi, D., & Jee, H. (2005). 3D welding and milling: Part I-a direct approach for freeform fabrication of metallic prototypes. *International Journal of Machine Tools and Manufacture, 45*, 1057–1062. https://doi.org/10.1016/j.ijmachtools.2004.11.021
28. Williams, S. W., Martina, F., Addison, A. C., Ding, J., Pardal, G., Colegrove, P., Martina, F., Addison, A. C., Ding, J., Pardal, G., Colegrove, P., Williams, S. W., Martina, F., Addison, A. C., Ding, J., Pardal, G., & Colegrove, P. (2016). Wire + arc additive manufacturing. *Materials Science and Technology*, 1–7. https://doi.org/10.1179/1743284715Y.0000000073
29. Martina, A. F., Ding, J., Williams, S., Caballero, A., & Quintino, L. (2019). Tandem metal inert gas process for high productivity wire arc additive manufacturing in stainless steel. *Additive Manufacturing, 25*, 545–550. https://doi.org/10.1016/j.addma.2018.11.022
30. Ding, D., Pan, Z., Van Duin, S., Li, H., & Shen, C. (2016). Fabricating superior NiAl bronze components through wire arc additive manufacturing. *Materials (Basel), 9*, 652. https://doi.org/10.3390/ma9080652
31. Rodrigues, T. A., Duarte, V., Miranda, R. M., Santos, T. G., & Oliveira, J. P. (2019). Current status and perspectives on wire and arc additive manufacturing (WAAM). *Materials (Basel), 12*, 1121. https://doi.org/10.3390/ma12071121
32. Selvi, S., Vishvaksenan, A., & Rajasekar, E. (2018). Cold metal transfer (CMT) technology - An overview. *Defence Technology, 14*, 28–44. https://doi.org/10.1016/j.dt.2017.08.002
33. Cong, B., Ding, J., & Williams, S. (2015). Effect of arc mode in cold metal transfer process on porosity of additively manufactured Al-6.3%Cu alloy. *International Journal of Advanced Manufacturing Technology, 76*, 1593–1606. https://doi.org/10.1007/s00170-014-6346-x
34. Somashekara, M. A., Naveenkumar, M., Kumar, A., Viswanath, C., & Simhambhatla, S. (2017). Investigations into effect of weld-deposition pattern on residual stress evolution for metallic additive manufacturing. *International Journal of Advanced Manufacturing Technology, 90*, 2009–2025. https://doi.org/10.1007/s00170-016-9510-7
35. Yang, D., He, C., & Zhang, G. (2016). Forming characteristics of thin-wall steel parts by double electrode GMAW based additive manufacturing. *Journal of Materials Processing Technology, 227*, 153–160. https://doi.org/10.1016/j.jmatprotec.2015.08.021
36. Oliveira, J. P., Barbosa, D., Fernandes, F. M. B., & Miranda, R. M. (2016). Tungsten inert gas (TIG) welding of Ni-rich NiTi plates: Functional behavior. *Smart Materials and Structures, 25*, 03LT01. https://doi.org/10.1088/0964-1726/25/3/03LT01
37. Pan, Z., Ding, D., Wu, B., Cuiuri, D., Li, H., & Norrish, J. (2018). Arc welding processes for additive manufacturing: A review. In Q. Han, Y. Li, & G. Zhang (Eds.), *Transactions on intelligent welding manufacturing* (pp. 85–93). https://doi.org/10.1007/978-981-10-5355-9
38. Oliveira, J. P., Fernandes, F. M. B., Miranda, R. M., Schell, N., & Ocaña, J. L. (2016). Materials characterization effect of laser welding parameters on the austenite and martensite phase fractions of NiTi. *Materials Characterization, 119*, 148–151. https://doi.org/10.1016/j.matchar.2016.08.001
39. Wang, F., Williams, S., & Rush, M. (2011). Morphology investigation on direct current pulsed gas tungsten arc welded additive layer manufactured Ti6Al4V alloy. *International Journal of Advanced Manufacturing Technology, 57*, 597–603. https://doi.org/10.1007/s00170-011-3299-1
40. Jhavar, S., Jain, N. K., & Paul, C. P. (2014). Development of micro-plasma transferred arc (μ-PTA) wire deposition process for additive layer manufacturing applications. *Journal of Materials Processing Technology, 214*, 1102–1110. https://doi.org/10.1016/j.jmatprotec.2013.12.016

41. Martina, F., Mehnen, J., Williams, S. W., Colegrove, P., & Wang, F. (2012). Investigation of the benefits of plasma deposition for the additive layer manufacture of Ti-6Al-4V. *Journal of Materials Processing Technology, 212*, 1377–1386. https://doi.org/10.1016/j.jmatprotec.2012.02.002
42. Gu, J., Ding, J., Williams, S. W., Gu, H., Ma, P., & Zhai, Y. (2016). The effect of inter-layer cold working and post-deposition heat treatment on porosity in additively manufactured aluminum alloys. *Journal of Materials Processing Technology, 230*, 26–34.
43. Martina, F., Colegrove, P. A., Williams, S. W., & Meyer, J. (2015). Microstructure of interpass rolled wire + arc additive manufacturing Ti-6Al-4V components. *Metallurgical and Materials Transactions A, Physical Metallurgy and Materials Science, 46*, 6103–6118. https://doi.org/10.1007/s11661-015-3172-1
44. Colegrove, P. A., Coules, H. E., Fairman, J., Martina, F., Kashoob, T., Mamash, H., & Cozzolino, L. D. (2013). Microstructure and residual stress improvement in wire and arc additively manufactured parts through high-pressure rolling. *Journal of Materials Processing Technology, 213*, 1782–1791. https://doi.org/10.1016/j.jmatprotec.2013.04.012
45. Xie, Y., Zhang, H., & Zhou, F. (2016). Improvement in geometrical accuracy and mechanical property for arc-based additive manufacturing using metamorphic rolling mechanism. *Journal of Manufacturing Science and Engineering, 138*, 1–8. https://doi.org/10.1115/1.4032079
46. Hönnige, J. R., Colegrove, P., & Williams, S. (2017). Improvement of microstructure and mechanical properties in wire + arc additively manufactured Ti-6Al-4V with machine hammer peening. *Procedia Engineering, 216*, 8–17. https://doi.org/10.1016/j.proeng.2018.02.083
47. Sun, R., Li, L., Zhu, Y., Guo, W., Peng, P., Cong, B., Sun, J., Che, Z., Li, B., Guo, C., & Liu, L. (2018). Microstructure, residual stress and tensile properties control of wire-arc additive manufactured 2319 aluminum alloy with laser shock peening. *Journal of Alloys and Compounds, 747*, 255–265. https://doi.org/10.1016/j.jallcom.2018.02.353
48. Baufeld, B., Brandl, E., & Van Der Biest, O. (2011). Wire based additive layer manufacturing: Comparison of microstructure and mechanical properties of Ti-6Al-4V components fabricated by laser-beam deposition and shaped metal deposition. *Journal of Materials Processing Technology, 211*, 1146–1158. https://doi.org/10.1016/j.jmatprotec.2011.01.018
49. Montevecchi, F., Venturini, G., Grossi, N., Scippa, A., & Campatelli, G. (2018). Idle time selection for wire-arc additive manufacturing: A finite element- based technique. *Additive Manufacturing, 21*, 479–486. https://doi.org/10.1016/j.addma.2018.01.007
50. Wu, B., Pan, Z., Ding, D., Cuiuri, D., Li, H., & Fei, Z. (2018). The effects of forced interpass cooling on the material properties of wire arc additively manufactured Ti6Al4V alloy. *Journal of Materials Processing Technology, 258*, 97–105. https://doi.org/10.1016/j.jmatprotec.2018.03.024
51. Mukherjee, T., Zhang, W., & DebRoy, T. (2017). An improved prediction of residual stresses and distortion in additive manufacturing. *Computational Materials Science, 126*, 360–372. https://doi.org/10.1016/j.commatsci.2016.10.003
52. Busachi, A., Erkoyuncu, J., Colegrove, P., Martina, F., & Ding, J. (2015). Designing a WAAM based manufacturing system for defence applications. *Procedia CIRP, 37*, 48–53. https://doi.org/10.1016/j.procir.2015.08.085
53. Sames, W. J., Medina, F., Peter, W. H., Babu, S. S., & Dehoff, R. R. (2014). Effect of process control and powder quality on Inconel 718 produced using electron beam melting. In *8th international symposium on superalloy 718 and derivatives* (pp. 409–423). https://doi.org/10.1002/9781119016854.ch32
54. Bai, J., Ding, H. L., Gu, J. L., Wang, X. S., & Qiu, H. (2017). Porosity evolution in additively manufactured aluminium alloy during high temperature exposure. In *IOP conference series: Materials science and engineering* (Vol. 167, p. 012045). IOP Publ. https://doi.org/10.1088/1742-6596/755/1/011001
55. Wang, L., Xue, J., & Wang, Q. (2019). Correlation between arc mode, microstructure, and mechanical properties during wire arc additive manufacturing of 316L stainless steel. *Materials Science and Engineering A, 751*, 183–190. https://doi.org/10.1016/j.msea.2019.02.078

56. Zeng, Z., Cong, B. Q., Oliveira, J. P., Ke, W. C., Schell, N., Peng, B., Qi, Z. W., Ge, F. G., Zhang, W., & Ao, S. S. (2020). Wire and arc additive manufacturing of a Ni-rich NiTi shape memory alloy: Microstructure and mechanical properties. *Additive Manufacturing, 32*. https://doi.org/10.1016/j.addma.2020.101051
57. Mcandrew, A. R., Alvarez, M., Colegrove, P. A., Hönnige, J. R., Ho, A., Fayolle, R., Eyitayo, K., Stan, I., Sukrongpang, P., Crochemore, A., & Pinter, Z. (2018). Interpass rolling of Ti-6Al-4V wire + arc additively manufactured features for microstructural re fi nement. *Additive Manufacturing, 21*, 340–349. https://doi.org/10.1016/j.addma.2018.03.006

Correction to: Investigation of Mechanical Properties in Friction Stir Welded Mg AZ 31 Alloy Workpieces

N. Giri, G. S. Brar, and A. S. Shahi

Correction to:
Chapter 7 in: C. Prakash et al. (eds.), *Additive, Subtractive, and Hybrid Technologies*, Mechanical Engineering Series, https://doi.org/10.1007/978-3-030-99569-0_7

The published version of the book has incorrect affiliations of the chapter authors:

The correct affiliations are as follows:

1. G. S. Brar—Department of Mechanical Engineering, National Institute of Technology, Srinagar, Uttarakhand, India

2. A. S. Shahi—Department of Mechanical Engineering, Sant Longowal Institute of Engineering and Technology, Sangrur, Punjab, India

This error has been corrected, and the author's affiliations were included in the book.

The updated version of this chapter can be found at
https://doi.org/10.1007/978-3-030-99569-0_7

C. Prakash et al. (eds.), *Additive, Subtractive, and Hybrid Technologies*,
Mechanical Engineering Series, https://doi.org/10.1007/978-3-030-99569-0_14

Index

C. Prakash et al. (eds.), *Additive, Subtractive, and Hybrid Technologies*, Mechanical Engineering Series, https://doi.org/10.1007/978-3-030-99569-0

T

U

V

The manufacturer's authorised representative in the EU is Springer Nature Customer Service Centre GmbH, Europaplatz 3, 69115 Heidelberg, Germany. If you have any concerns regarding our products, please contact ProductSafety@springernature.com

Printed and bound by CPI Group (UK) Ltd, Croydon, CR0 4YY
30/09/2026
02225265-0016